Quality Control – Intelligent Manufacturing, Robust Design and Charts

Edited by Pengzhong Li, Paulo António Rodrigues Pereira and Helena Navas

Published in London, United Kingdom

IntechOpen

Supporting open minds since 2005

Quality Control - Intelligent Manufacturing, Robust Design and Charts
http://dx.doi.org/10.5772/intechopen.87736
Edited by Pengzhong Li, Paulo António Rodrigues Pereira and Helena Navas

Contributors
Ross Stewart Sparks, Hossein Hazrati-Marangaloo, Alexis Oliva, Matías Llabrés, Ioannis S. Triantafyllou, Shadha ALAmri, Ercan Oztemel, Emmanuel Karabo Mpodi, Tjiparuro Zeundjua, Oduetse Matsebe, Amusan Lekan, Clinton Aigbavboa, Moses Eterigho Emetere, James Owolabi, Joseph Evans Agolla, Anouar Acheghaf, Naima Amar Touhami, Coşkun Hamzaçebi, Asif Ahmad, Helena V. G. Navas, Raul Navas, Sónia R. Bentes, Fatma Hentati, Neila Masmoudi

First published in London, United Kingdom, 2021 by IntechOpen
IntechOpen is the global imprint of INTECHOPEN LIMITED, registered in England and Wales, registration number: 11086078, 5 Princes Gate Court, London, SW7 2QJ, United Kingdom
Printed in Croatia

British Library Cataloguing-in-Publication Data
A catalogue record for this book is available from the British Library

Additional hard and PDF copies can be obtained from orders@intechopen.com

Quality Control - Intelligent Manufacturing, Robust Design and Charts
Edited by Pengzhong Li, Paulo António Rodrigues Pereira and Helena Navas
p. cm.
Print ISBN 978-1-83962-497-1
Online ISBN 978-1-83962-498-8
eBook (PDF) ISBN 978-1-83962-499-5

For EU product safety concerns:
IN TECH d.o.o., Prolaz Marije Krucifikse Kozulić 3, 51000 Rijeka, Croatia, info@intechopen.com or visit our website at intechopen.com.

Meet the editors

Dr. Pengzhong Li is a professor in the School of Mechanical Engineering and a teaching director in the Sino-German School of Postgraduate Studies (CDHK), Tongji University, Shanghai, China. He received his Ph.D. degree in Mechanical Engineering from Tongji University in 2004. From 1995 to 2001, he was the manager in the Business and Warehousing Management Departments of Guilin Daewoo Bus Co., LTD. He is a Director of Intelligent Manufacturing and Services Branch of China Creative Studies Institute and Director of East China Branch of National College Institute of Manufacturing Automation. Dr. Li's research interests include intelligent manufacturing systems and quality control technology in manufacturing. At present, he is committed to research on man–machine coordination in the application of artificial intelligence.

Dr. Pereira received his Ph.D. in Biotechnology from the Catholic University of Portugal. He has been recruited as a quality and laboratory expert for seminars and professional laboratory meetings throughout Europe, Africa, and South America. He has more than twenty-five years of experience working in medical laboratories, having held key scientific leadership roles: 15+ years as a senior researcher; 10+ years as a consultant for a metrology laboratory based on ISO/IEC 17025 specifications and related standards; 20+ years as a consultant and auditor of quality management systems based on ISO 9001, ISO/IEC 17025, and ISO 15189 standards; 16+ years as the quality manager in the Portuguese Institute of Blood and Transplantation, including more than 6 years in national coordination; and 6+ years as a professor of Quality Assurance. Currently, he is the head of the R&D Department at the Portuguese Institute of Blood and Transplantation, Lisbon, Portugal. Dr. Pereira is the author of several peer-reviewed scientific articles and indexed book chapters, and an editor for several books. He serves as a member of several editorial boards. He is a member of the Clinical Laboratory and Standards Institute and Eurachem. Dr. Pereira is an expert on in vitro diagnostic medical devices at the European Commission.

Helena Navas is a professor in the Department of Mechanical and Industrial Engineering, NOVA School of Science and Technology (FCT NOVA), Portugal, and a Researcher for the Unit for Research and Development in Mechanical and Industrial Engineering (UNIDEMI). She represents the Portuguese Association for Quality (APQ) in the Portuguese Technical Standardization Committee on Research Activities, Development, and Innovation (RDI) and represents FCT NOVA in the Portuguese Standardization Committee on Project Management. She is a member of the Scientific Council of the Portuguese Society of Metrology (SPMet). She was a member of the International Executive Board of the European TRIZ Association (ETRIA) for the 2011–2012 term. She has been a columnist for the Innovation and Entrepreneurship newsletter since

April 2014. She is a lean leader and completed the Six Sigma Black Belt Manufacturing course from The Lean Six Sigma Company. Professor Navas has been a guest speaker at several seminars, round tables, and events dedicated to innovation. She is a researcher, consultant, and trainer in innovation, systematic innovation, TRIZ methodology, and lean and continuous improvement.

Contents

Preface

The rise of artificial intelligence, the Internet, and other emerging technologies has promoted the deep integration of new information and advanced communication and manufacturing technologies. It has also encouraged the transformation of manufacturing and services to digitalization, networking, and intelligence, which is undoubtedly a revolution in manufacturing methods and technical means. This transformation inevitably affects all aspects of the strategic orientation, resource allocation, and management of enterprises.

As one of the key factors of enterprise management, quality control will also change along with changes in the manufacturing environment and factors affecting product quality (man, machine, material, method, and environment). A series of changes will occur in management contents, methods, capabilities, and real-time management effectiveness and efficiency. Some of these changes are revolutionary.

The several following features will have a significant impact on quality management.

Individual customer needs and rapid response. Identifying customer needs is the first step in quality management. Customer needs will gradually change from small varieties and large batches to multiple varieties and small batches; individualized and differentiated customer needs will become the norm. The basic quality management concept of "customer-focused" will be further strengthened with rapid response to customer needs and continuous satisfaction.

Customer participation in product design and development. The interconnection in the era of intelligent manufacturing provides the technical means for customers to participate in product design and development. Customers can participate in reviewing, verifying, and validating product design and development, as well as product type testing, reliability testing, and other activities either remotely or on-site, providing the technical means to fully implement customers' individual needs.

Further improved process capability. The technical means and control capabilities of process control are continuously improved to make production more convenient and flexible. Process management based on big data is more accurate, optimized, and mature, so that the predictive maintenance of equipment is gradually realized. Product identification, traceability, and process error prevention are more thorough through technical means such as electronic labels and QR codes. Quality monitoring (including statistical process control) is a real-time process, and quality tools and methods are convenient to use.

Easy-to-use quality management tools and methods based on statistical technology. Quality management tools based on mathematical and statistical technology, such as the seven basic quality tools, statistical process control (SPC), and others, involve a large amount of data statistics and calculations. As such, an intelligent manufacturing system will automatically collect, analyze, and share this data, which needs to be collected in real-time according to different conditions set in advance, making the use of quality management tools more convenient.

The fields of traditional manufacturing, which were once well defined and distinct, have been completely overturned by new technologies, and each field extends, covers, and overlaps with the other. While intelligent technologies and digitalization strategies have brought about big changes in the manufacturing industry, quality control remains an important area that needs close attention. If a manufacturer has completed the intelligent transformation in design, production, testing, packaging, and other processes, but has not updated the quality control system accordingly, there will be a big gap between the current process and the product vision. This will ultimately lead to failure to realize the product and service promises made to customers. To optimize and upgrade the entire manufacturing process, quality control methods must be simultaneously optimized as well.

Therefore, quality control is an essential factor in the era of intelligent manufacturing. Real and comprehensive innovations are needed and, fortunately, many researchers and institutes have developed several innovative technologies to optimize existing quality control systems and manage complex and mass production details.

Focused on new trends and developments in quality control from a worldwide perspective, this book presents the latest results on novel approaches to current problems in quality control. Written by academicians, researchers, and practicing engineers, this volume contains thirteen chapters organized around three topics: intelligent manufacturing, robust design, and control charts. The information contained herein is useful for technical experts and entrepreneurs.

This book would not have been possible without the generous assistance of many colleagues. The authors and I would like to express our sincere appreciation to my co-editors, Professor Helena Navas and Dr. Paulo Pereira. We would also like to thank Ms. Anja Filipovic and Ms. Jasna Bozic at IntechOpen for their patient and careful organization and promotion in the publication of this book.

Pengzhong, LI
Tongji University (School of Mechanical Engineering; Chinesisch-Deutsches Hochschulkolleg),
Shanghai, China

Paulo António Rodrigues Pereira
Portuguese Institute of Blood and Transplantation,
Portugal

Helena Navas
Universidade Nova de Lisboa,
Portugal

Chapter 1

Introduction to Intelligent Quality Management

Ercan Oztemel

Abstract

Intelligent manufacturing is becoming more and more attractive for industrial societies especially after the introduction of industry 4.0 where most of industrial operations are to be carried by robots equipped with intelligent capabilities. This explicitly implies that the manufacturing systems will entirely be integrated and all manufacturing functions including quality control and management will have to be made as much intelligent as possible in operating with minimum human intervention. This Chapter will present a brief overview of some implications about intelligent quality systems. It intends to provide the readers of the book to understand how the concept of artificial intelligence is to be embedded into quality functions. It is known that the interoperability is the rapid transformation requirement of industry specific operations. This requires the integration of quality functions to other manufacturing functions for sharing the quality related knowledge with other manufacturing functions in order to sustain total intelligent collaboration. Achieving this, on the other hand, ensures the improvement of manufacturing processes for better performance in an integrated manner. Note that, although some general information about intelligent manufacturing systems are given, this chapter is particularly focused on discussing intelligent quality related issues.

Keywords: intelligent quality, intelligent manufacturing, integrated quality, quality management, quality improvements

1. Introduction

Intelligent manufacturing is becoming more and more attractive for industrial societies especially after the emergence of industry 4.0 where most of industrial operations are to be carried by robots equipped with intelligent capabilities. Since digital transformation is increasing every day. The manufacturing societies are enforced not only to increase the development speed of manufacturing systems but also to improve the functionality, flexibility, usability and interoperability of the system developed.

For each manufacturing function, different methods and methodologies are developed to sustain the level of intelligence due to the nature of the operations carried out within the scope. This applies to quality operations as well. Introducing artificial intelligence in quality systems were considered since so many years. Pham and Oztemel (1996) Published a book on intelligent quality systems and took the attention of both research and industrial communities on this issue [1]. They presented real life applications and highlighted possible areas of future developments. Oztemel and Tekez (2008) took this initiative one step ahead and proposed a multi agent quality management system where each quality function is considered to be self-operating and capable of working independently but coordinating with others in a well-designed manufacturing suits [2]. Their approach contributed to the developed a reference model of fully integrated, intelligent manufacturing system so called REMIMS [3]. Application of artificial intelligence techniques and methodologies to quality domain varies. There are vast amount of research and literature available. Here in this Chapter, only a few of them (that are different nature in a wide spectrum) will be mentioned for indicating the importance of this initiatives and the range of applications.

Computer vision technology and artificial intelligence, for example, are well utilized for quality control purpose in many different applications. As the quality of many products in manufacturing is defined by the dimensions and surface features, computer vision technology is utilized for mainly replacing human eyes. This attracted manufacturing community due to cost effectiveness. It is now well proven that together with utilizing statistical analysis methods, automated machine vision systems are capable of analyzing geometric and surface features for deciding about the quality of products [4].

Similarly, Xu et al. (2018) provided an intelligent quality problem solving system capable of performing a comprehensive analysis of data mining process and methods. They run a pilot application in automotive manufacturing company to demonstrate intelligent capabilities of the system proposed [5]. Bihi et al. (2018) designed and intelligent quality management systems which can learn and adapt itself to the requirements of flexible manufacturing systems [6]. The goal of this system is to learn line layout and setup through predefined training cycles. Tseng et al. (2016) presented a remote part tracking and quality control system, so called e-quality control, based on the application of support vector machine (SVM) to predict the output quality [7].

Like some of those mentioned above, a comprehensive literature review will provide huge amount of information on implementing artificial intelligence on quality functions. The information stated above is considered to be sufficient enough for taking the attention of the reader to intelligent quality operations. Note that, the remaining part of this chapter will highlight the importance of intelligent tools for sustaining the quality of systems, processes and products. Also note that, the Chapter will describe a general framework for utilizing intelligent quality systems.

2. Introducing intelligence to manufacturing functions

Success of manufacturing systems in industry 4.0 era will be heavily dependent upon their capability to perform operations with a certain degree of autonomy which can be assured by employing artificial intelligence in designing manufacturing systems to some extent. Oztemel (2010) describes a general framework with possible characteristics of intelligent manufacturing systems [8]. Due to wide variety of products and services requiring very complex operations, manufacturing systems are facing so many challenges. For example, customization is a very demanding properties of the products and services for intended customers. There seems to be several unique requirements like this one for the manufacturing systems to cope with the existing transformation need. Some of these are;

- to be able to adopt new models,
- generating new forms of manufacturing,
- implementing new methodologies for achieving the required level of transformation to smartness in the traditional manufacturing systems.

It is now very obvious that generating a required level of flexibility in manufacturing system is not enough for proper transformation. Whole supply chain is to be involved. Unmanned operations and systems, on the other hand, generates more competitiveness and puts more challenges for those which are having lack of knowledge on generating the required autonomy. Since, artificial intelligence, information technology, and robots with well performing sensors are the technologies to form advanced manufacturing systems which are to be capable of overcoming those challenges, manufacturers should spend a great amount of effort to learn and experiment these in their operations. In other words, the manufacturers may need to follow lean principles together with intelligent equipment. This means, they need to optimize productivity and effectiveness of their manufacturing hardware, reduce possible wastes (scraps, overtimes, costs etc.), reduce cycle times and increase the expected output.

Intelligent equipment should not only be able to utilize artificial intelligence and machine learning technologies, but also be capable of processing big data collected over a certain amount of sensors in different roles for the sake of sustaining proactive management of equipment, processes services as well as products. Generating intelligent systems in this manner requires utilizing advanced information and manufacturing technologies for sustaining intelligence, re-configurability, interoperability, reusability and flexibility. Shen and Norrie (1999) provided a survey pointing out various requirements of especially agent based intelligent manufacturing systems [9]. Note that intelligent manufacturing, requires certain type of technologies for assuring the manufacturing equipment to behave as much intelligent as possible. Internet of Things, Cyber-physical systems, cloud computing, big data analytics, learning events, generating immediate responses to unexpected changes around etc. will be inevitable to sustain required level of smartness. Note that the production life cycle as a whole should be integrated and motivated by AI enriched sensors, decision making systems with big data analytics as well as advanced materials [10]. It seems that the competitiveness of manufacturing systems will heavily be dependent upon the level of intelligence they possess. Especially, utilizing data integration capabilities of information systems, intelligent decision making and reasoning capability (cognitive evaluation) over the data available, representing the results of the cognition process through dashboards and interactive visual analytics, employing smart sensor technologies so on and so forth will be the main focus of manufacturers in the following decade.

For a manufacturing system, being able to process real time data gathered from the machines and performing intelligent analysis over those through implementing AI technologies make it to predict and understand critical events and solve problems immediately before yielding any dangerous and hazardous situation as well as wastes. This capability of manufacturing systems even allows the machine to perform predictive maintenance and generate well operating manufacturing suits.

Current literature is rich enough to provide information regarding intelligent manufacturing as well as various applications. The reader may not have any difficulty in reaching those. For the purpose of this Chapter, it would be beneficial to mention some of the areas where AI technology is heavily utilized in generating intelligent behavior of certain type of quality functions.

3. The need for integrated manufacturing system

Since human intervention is minimized in intelligent operations, the systems should be capable of communicating with one another in order to make sure that the full manufacturing cycle is to operate without and delays and problems. This implies that manufacturing systems should employ knowledge protocols or any other knowledge transfer language [11, 12]. The integrated architecture in this manner ensures the detection of undesired operations and enables generation of immediate responses. This, in turn, improves the quality, minimizes downtime and improves overall equipment efficiency. Through digital operations and simulation technology (empowering digital twins) allow to virtually experiment manufacturing operations and realize possible shortcomings which could be prevented before actual set up.

Smart operations also empower manufacturing systems in terms of optimizing the overall supply chain. Operations like demand forecasting, inventory optimization, supplier monitoring and performance assessments can be made more effective through intelligent methods and internet of things with equipped sensors.

Visual system manipulations, on the other hand, seem to be more and more demanding in manufacturing sites especially after the introduction of 5G. Being able to experiment manufacturing operations in virtual world and generate digital replicates of the manufacturing operations, make the life easy for designers to be able to discover and foresee the effect of new business models, new technology implementations and smart decision making capabilities.

It should be noted that the technological progress and disruptive technologies improve competitive advantage and makes it possible to devices systems which are superior over traditional automation systems. Intelligent manufacturing operations and the capability of real time data processing, on the other hand, allows prediction of possible failures before they actual occur and highlights possible anomalies. This may encourage predictive maintenance and prevent downtime and machine failures. With this respect, the quality management in integrated manufacturing environment is essential in order to sustain competitive manufacturing.

4. A general framework for intelligent quality

Among various other applications, smart systems can be used to assure process and product quality. One of the distinguishing character of intelligent quality in recent years is that, not only the quality of the productions processes is the subject of quality management but also all processes within the production life cycle of the enterprise including the quality of procurement, design, planning, production, distribution and marketing. The applications can therefore be grouped into 3 categories as intelligent quality implementations before, during and after manufacturing operations. To mention some of those, smart manufacturing systems are to be capable of assuring quality through experimental design before production process starts [13]. During the production process some statistical process control (SPC) systems ca be well effectively employed thorough intelligent analysis and implementations [14]. Similarly defect detection after production is subject to intelligent operations. Similarly, visual inspection technology based on computer vision and machine learning is not strange in quality management societies. Wang et al. (2019) presented a computer vision based inspection system by employing deep learning in order to identify and classify defective products without the loss of accuracy [15]. Chesalin et al. (2020) provided an overview of intelligent quality tools that can be utilized for handling quality related issues within a manufacturing suit [16].

Similarly, reviewing respective literature highlights huge amount of areas where intelligent and smart systems are being utilized for the sake of generating a better quality on processes, products and services provided. Some of them can be listed below.

- Sustaining the quality of processes, products and services, ensuring reliability and detecting possible anomalies.
- Generating alerts for process abnormalities and fault detection.
- Predicting the behavior of the machines, devices and respective equipment in terms of expected yield.
- Maintaining machine and condition monitoring for making sure that the machines are healthy enough to operate.
- Assuring the effectiveness of whole supply chain from suppliers to the customers including managing the respective resources as well as customer segmentation.
- Handling inventory and optimizing material management.
- Performing computer vision based inspection through implementing machine learning techniques in order to find out defective products
- Implementing intelligent systems to optimize quality parameters of the process (experimental design)
- Benchmarking which would allow the systems to be compared with their counterparts.

Considering the complexity and functionality of products and process and the need for handling huge amount of data, sustaining required level of quality is nearly impossible without taking the advantages of information technologies. Enriching those with intelligent capabilities (utilizing artificial intelligence) and data analytics (big data manipulation) definitely improve the performance of quality management activities.

Another aspect of intelligent quality in modern manufacturing systems is to sustain continuous improvement as was deployed by human quality operators and process teams. This definitely require communication of various manufacturing agents designed to exchange required knowledge which can trigger improvement activities.

As above explanations clearly imply, well known artificial intelligence methodologies such as expert systems, machine learning (neural networks), genetic algorithms, fuzzy logic and some others are proven to be good enough for generating solutions to respective quality situations. However, for some situations these methodologies may not generate required responses alone. It may be inevitable to utilize more than one of these. Intelligent agents are capable of utilizing various AI methodologies depending upon the nature of the problem. The studies experimented indicates more effective and solution providers to the manufacturers.

4.1 Requirements for developing intelligent quality management systems

Intelligent quality requires management commitment in order to provide necessary facilities and capabilities for complying with the requirements and continually improve the effectiveness of the quality system. It should allow quality responsible to be able to assess the status of quality objectives. However, not only the quality people be involved but all should take part in developing and running the systems. Everybody in manufacturing chain should aim continuous and sustainable quality level for enterprise wide operations and services. This ensures the required level of integration (quality functions with all others). Meaning of this is to communicate quality problems and protect the other systems to be negatively affected while improving quality culture. Having the commitment of all people also try to enable quality systems to support overall business objectives as set forth by the top management.

Another aspect of intelligent quality is to employ a knowledge driven approach for generating and operating the systems as opposed to traditional data driven one. In order to establish required level of intelligence the system should be equipped with respective operational as well as respective quality knowledge. Traditional data based approaches are not sufficient enough to generate intelligent behavior. There is a need for methods and methodologies to dig out knowledge out of available data (this may be called as knowledge mining) for generating self-behavior of intended quality function. Artificial intelligence and machine learning systems provide various alternatives for locating and utilizing quality knowledge.

Due to the nature of intelligent systems, when they are well defined, they provide quality know-how for those which are having difficulty in setting up quality systems. Since they will be equipped with relate domain and quality knowledge, they would prevent;

- lack of quality in process outputs automatically by taking the attention of the operators or other systems to anomalies and defective outputs. They may also recommend some solutions and remedies.

- lack of quality knowledge as well as related process knowledge to some extent.

- lack of experience in handling quality related issues especially for those who are new in practicing quality operations

- poor system development through highlighting missing parts and required capabilities.

- lack of methods and methodologies to be implemented.

- misleading machine set up and respective inefficiency.

- misleading knowledge and process flow operations.

- inaccurate interfaces between processes and operational units.

Moreover, being able to generate a good quality product and services by intelligent quality practices increase customer trust and loyalty. Since the systems will have to self-operate and make recommendations and remedies for the responsible people, they may also find the system attractive for their success.

With this understanding intelligent quality should aim to satisfy following requirements. These may be called as “the baseline requirements” for intelligent quality management systems. Achieving these, verify, validate and assure the effectiveness of the systems employed.

- Minimizing duplication of work and various types of functionalities.
- Gathering, updating and storing the real time data and making it timely available.
- Generating quality knowledge out of data available and using those for self-decision making.
- Making accurate predictions about the status of manufacturing equipment as well as expected outputs (defect free deployment)
- Assessing supplier automatically and monitoring their performance through on-line tracking systems.
- Generating alerts for the responsible people to take immediate actions where ever the system is not able to do so.
- Measuring and sustaining the reduction of the following by means of digital infrastructure.
    - Customer (both internal and external) complaints
    - Defective products and scrap rates
    - Process anomalies
    - Delayed orders
    - Downtime due to machine unavailability.
- Measuring and sustaining the increase of the following by continuous monitoring systems.
    - Productivity in manufacturing lines
    - Employee satisfaction
    - Customer (both internal and external) satisfaction
    - Supplier satisfaction
- Maintaining quality standards as they evolve and being adaptive to cope with the changes without spending too much effort. That is to adaptively comply with quality regulations.
- Sustaining data security as well as information integrity.

4.2 Framework for intelligent quality systems

Above explanations indicate the importance of intelligent manufacturing in general and intelligent quality in specific. This section provides a general framework for generating an integrated intelligent quality system. As implied above, main requirement behind this framework is to set up knowledge intensive activities for assuring quality and continuous improvement. This, in turn, requires a systematic performance monitoring and assessment. The proposed framework is based on achieving business objectives which in turn needs to be aligned with the objectives of the processes. Note that, intelligent tools can support quality management activities in various ways as listed below.

- Handling quality policies and objectives of any manufacturing system could be supported by intelligent tools in order to set up proper quality plans and standards.

- Quality management related knowledge could very well be stored in knowledge bases and utilized in making quality decisions. The performance of the system will be depended upon the acquisition and application of relevant knowledge sets.

- Data from various sensors could be gathered and interpreted for the sake of sustaining on-line monitoring of the process in terms of quality requirements.

- Design and operational reliability (ability to perform a specific function) availability (ability to keep a functioning state) maintainability (ability to be timely and easily maintained) and safety (ability not to harm people, the environment, or any assets) analysis (SARM) could be carried out intelligently for assuring engineering characteristics of products or systems. SARM is proven to be good in identifying, analyzing, evaluating, preventing, verifying and correcting the defects and hazards of any system. Intelligent tools could facilitate these analyses more effectively and efficiently for the benefit of the enterprise.

Taking above explanation into account, it would be possible to say that intelligent quality management operations are carried out in 3 different phases of any manufacturing system. These are;

- Phase-1: Operations before running manufacturing processes

- Phase-2: Operations during running manufacturing processes.

- Phase-3: Quality Operations after manufacturing

All of these phases should be supported by intelligent tools in various nature and should be involved within given framework for the sake of integration and completeness of the whole life cycle of quality management activities.

4.2.1 Quality operations before running manufacturing processes

Figure 1 indicates the basic elements of intelligent quality management in the first phase.

There are two aspects of specification analysis like process specification and product specification. These should be compliant with one another. Intelligent systems should be able to align and compare predefined set of requirements (specification match) and intelligently analyze those to generate complete set of specifications which could be used for properly designing the processes to run defined process plans. Fuzzy logic and expert systems could very well be employed for this.

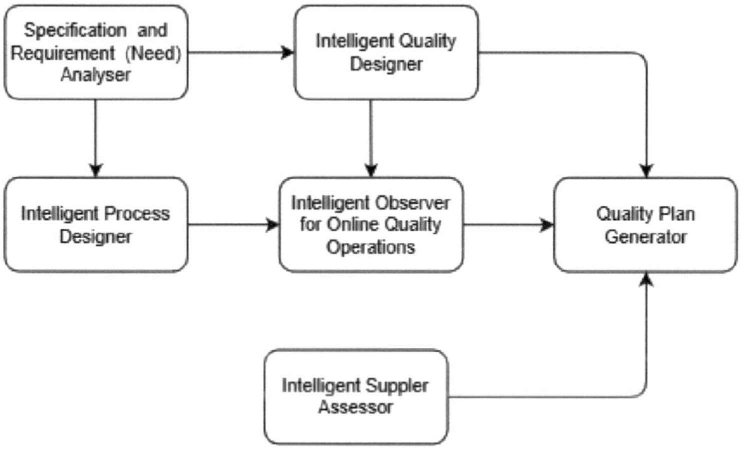

Figure 1.
Intelligent quality operations before manufacturing.

Agents equipped with machine learning systems can also help in refining process and product specifications and generate optimum designs [17].

Designing quality (design of experiment) is another area where machine learning and artificial intelligence can be utilized. Finding out the optimum values of the quality parameters would be possible by learning systems which can learn various parameters and help optimize the quality design [13].

Quality function deployment (QFD) is an effective tool to shape product specifications based on customer requirements. Supporting this function with intelligent tools may enable automatic prediction of product quality with respect to customer expectations [18]. This information would also be used in forming design and manufacturing data. When there is a change in customer demand (which is the case in most of time), the effect of that in manufacturing can be assessed. With this capability intelligent QFD may be used during manufacturing process as well. Adding new requirements, removing an existing one, changing the design attributes would be handled by the support of intelligent tools like fuzzy logic.

Supplier assessment is another area where intelligent systems would be employed. Fuzzy logic theory is heavily involved in selecting best suppliers among alternatives [19]. Similarly, literature shows evolutionary modeling [20], and neural network implementations [21]. This is indirectly related to quality operations as it ensures good quality and timely material with bearable costs to be available.

There are some information systems and software platforms for handling quality standards like ISO 9000 series. The tools which can facilitate especially corrective actions and run validation process over quality implementations. Computer based intelligent decision making along this line may generate effective implementation of quality standards. Enriching these systems with intelligent capabilities may empower quality assurance systems.

4.2.2 Quality operations during execution of manufacturing processes

Figure 2 indicates the basic elements of intelligent quality management in the second phase. As indicated, generating quality standards and quality assurance system is the baseline requirement for setting up intelligent quality operations during execution. This may assure the integration of quality operation for various manufacturing systems. As described by Albers et al. (2016), a comprehensive model can be devised in order to support both digitization process and sustaining

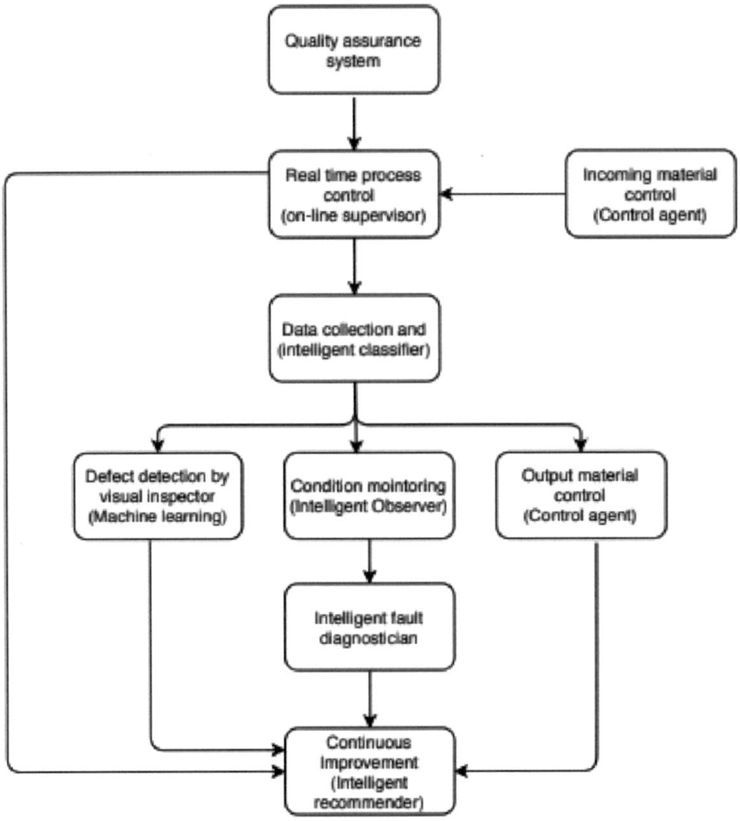

Figure 2.
Intelligent quality operations during manufacturing.

good quality in an integrated manner [22]. Similar studies indicate that, some of the methods and methodologies such visual and constructive inspection of products and processes, intelligent classifier and data acquisition systems, defect detection and fault diagnosis as well as condition monitoring systems are inevitable for real time quality management.

Making the data available through data gathering systems enables analysis and visualization of information which can be effectively used in generating evidences for quality related decision making. By detecting defects as early as possible in manufacturing line not only yields cost reduction but also eliminates causes of failures. This is in fact one of the main requirements for successful quality assurance systems.

Real time intelligent quality assurance may ensure that the smart operations are handle without interruption and sustain continuity of manufacturing operations with the required level of reliability, flexibility, and interoperability. This, in turn, increases productivity and effectiveness as well as to reduce the defects to the minimum level possible.

Real time and online quality operations handled by intelligent tools should support well-known quality tools such as quality function deployment, statistical process control, constructive and visual inspection, failure mode and effect analysis, measurement systems etc. Each of these tools could very well be enriched by self-behaving capability and reconfigured in accordance with the requirements and specification of manufacturing systems.

Statistical process control (SPC) is one of the major area for implementing artificial intelligence for sustaining good quality within a manufacturing process. Pham and Oztemel (1996) are long ago took the attention of the scientific community to this point by introducing an expert system (called XPC) and machine learning system for statistical process control [1]. XPC was able to construct quality control charts and perform process capability analysis in order to make sure that the process is able to comply with customer specifications, collect data on-line, plot those over the chart constructed and interpret the status of the process, perform fault detection and diagnosis, make some recommendation in case of an out of control situation. The system was also able to modify control limits and control chart used in order to embed improved process standards. It was well appreciated as the system was playing the role of a quality engineer and continuously monitoring the process on-line.

When utilizing machines, there is and will be a need for condition monitoring and try to identify possible machine faults before they cause unbearable costs. As described in Zabinski et al. (2019), condition monitoring systems mainly deals with, monitoring and feature extraction, real-time anomaly detection and fault diagnosis [23]. Expert systems and rule based reasoning mechanisms, neural networks and deep learning methods as well as intelligent agents can be extensively used for monitoring the machines and process for possible anomalies and faults. Literature provides huge amount of implementation in these areas. Besides, as condition monitoring is the best way to minimize the probability of failure and therefore maintenance cost, making it intelligent and generating a self-behaving observer can be an important tool for predictive maintenance [24].

By implementing AI techniques to monitor the condition of an equipment, identifying possible faults and observing the trend of the level of deterioration may allow possible actions to be taken at convenient times without interrupting production process. Generating intelligent predictive maintenance system by this way, may prevent major overhauls or increase the time to go into an overhaul. This definitely reduces significant damage that would otherwise be unavoidable. Above all, proper maintenance schedules can be generated in between the demands for having maintaining the equipment if not done automatically. Considering the amount of money spend for keeping spare parts of the machines in stock, intelligent condition monitoring may prevent undesired use of spare parts and avoid unnecessary parts not to be overstocked.

Artificial intelligent methods, especially machine learning techniques remarkably increased the effective utilization of visual inspection system for the sake of improving the quality of products and processes [25]. Note that, quality control through inspection is long being implemented in industry for locating nonconforming products or processes. Main aim is to deliver products and services to the customers in required specification (so called good quality) or defect free.

In earlier times, it was possible to inspect nonconformity by human eyes. Since the complexity and functionality of the systems are increased, it would not be possible to decide about the quality of the inspected items by human eyes. Computer support was inevitable and self-decision making which was assured by artificial intelligence was extremely useful. This is even more important today due to the nature of big data available within manufacturing areas. Two approaches are implemented.

- Taking the video record of the systems and products and then analyzing those off-line for finding if there is any abnormality.

- Real-time data collection and defect detection using image recognition and AI techniques such as neural networks.

Technology enables obtaining high quality video images with certain cameras specifically configured to manufacturing line. Data collected can be cleaned in real time for the sake of locating anomalies and defects such as scratch, dent, crack, dirt, wrong print, foreign objects, undesired bubbles etc. Computer vision system replicates human vision and assessment system.

Inspection systems may require both hardware (camera, gateway, photometer, colorimeter, CPU etc.) and software (video processor, database, reasoning mechanism and learning engine, cloud services etc.). Hardware is mainly used for data collection and processing whereas software is used for recognition and interpretation. In intelligent visual inspection, the learning and interpretation capability indicates the degree of smartness.

Production part approval process (PPAP) can be supported by intelligent tools for providing evidences that all customer demand-based specifications are well understood and proper manufacturing set up is in place. It may ensure that the manufacturing process is running in expected rate and producing well accepted products by the customers. It may also assure the fitness of quality plans, support maintaining design integrity, provide early detection of problems to be sorted, prevent nonconforming parts to be processed etc. Another benefit of intelligent PPAP is to make sure that the suppliers have well understood the engineering design attributes and specifications of the products they are vendoring.

Intelligent failure mode and effect analysis (FMEA) can be very useful for finding and diagnosing process and system faults. The system may handle the information of various failures and analyze and interpret those from multiple point of view. Characteristics of different failures within product and process life cycle can be well managed. As described by Zhao et al. (2004), the system may involve a model base and knowledge base as well as functional models of the products and effect analysis of possible failures [26]. It would be possible to perform various reasoning processes depending upon the type of failure in question. FMEA of very complex system can be carried out with implementing expert knowledge of the domain as well as respective failures. This system may be coupled with Fault Tress Analysis (FTA) which is a hierarchical representation of faults enabling the analysis for the probability of a failure and its consequences. Due to uncertainty, complexity and inexact information, it is not easy to define the probability of a fault to occur. However, intelligent tools such as fuzzy logic and neural network are very capable of handling imprecise knowledge [27]. Generating intelligent FMEA and FTA may allow generating records of failures and recommends some corrective actions for improving the quality of manufacturing systems.

Creating and integrating all of the systems, some of those are listed above, may lead fully automated intelligent quality system within a smart manufacturing environment. Being able to receive data and information from various manufacturing functions and related systems without any interruption enables the designers to device a recommendation system to prevent misleading operations and deterioration from respective manufacturing plans. Recommendation could be provided for;

- Foreseeing possible deteriorations
- Preventing errors and faults over the process or products
- Correcting errors and faults
- Preventing the shift of non-conformed products between processes.
- Preventing the delivery of non-conformed products to customers.

Some of the recommendations would be carried out by intelligent systems capable of self-improving whereas some of them would still be sorted by the human operators.

4.2.3 Quality operations after manufacturing

Upon completing manufacturing, the products, some quality operations need to be carried out in delivery and after sale services. There could be various activities some of those are given in **Figure 3**.

Performing quality predictions before and after manufacturing can be carried out intelligently in order to follow if the predicted quality can be realized or not. Intelligent post production analyzer receives not only the predicted values but also some warranty information to keep track of the products after they delivered to the customers. Products should comply with the specifications and terms of use as promised before order received. The system may define conditions subject to warranty as well.

Comprehensive quality analysis may also be carried out upon receiving some feedback from the customer. Feedback can be received either by satisfaction surveys and information gathered by customer information channels (i.e. on-line support through web page applications). Information provided by the customers may be a good source of updating quality standards and expectations within manufacturing systems.

Analyzing the quality costs may provide some useful information for quality related operations and their effectiveness. Cost benefit analysis could direct improvements over the quality management process. Deterioration and sampling costs as well as the cost of inspection and control systems for specific products with specific attributes/specifications. Intelligent cost analyzer may collect cost data and may establish true costing and pricing strategies which would not compromise from the quality but increase the attractiveness of the products and services. Analyzer may generate some recommendations for quality improvements.

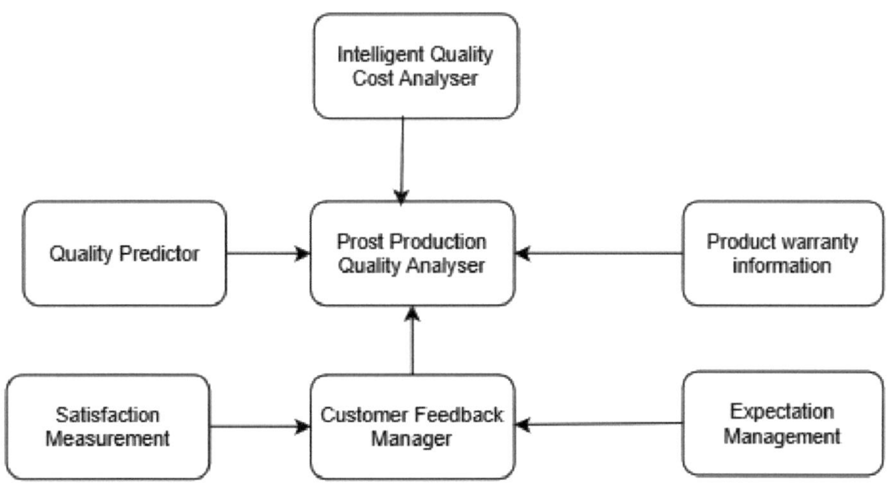

Figure 3.
Intelligent quality operations after manufacturing.

5. Conclusion

Intelligent quality management is attracting the attention of manufacturing society as it provides much opportunities to improve quality of both products and services. Agent based systems, knowledge based system, computer vision and intelligent inspection systems are proven to be capable of sustaining the quality of manufacturing systems as well as the quality of respective outputs.

As outlined in this Chapter, artificial intelligence and intelligent tools can be utilized for quality operations before, during and after manufacturing. Since each manufacturing proses is unique in terms of attributes, characteristics, and qualities products and processes. Special attention is to be given in developing intelligent quality systems. Relevant expertise and domain knowledge need to be acquired and presented to the computer for generating reasoning about the respective quality function.

Developing real-time monitoring and failure prediction systems, real time observation through visual inspection, data analysis and visualization of information which drives the evidence-based decision making and integration of the manufacturing systems throughout the whole supply chain and manufacturing life cycle could very well handled with the support of intelligent system generation technologies. This in turn supports continuous improvement of manufacturing systems in terms of reaching better quality.

The examples from the literature clearly indicates successful implementations. Sustainable quality can be better ensured with the support of intelligent tools. The manufacturers may have so many opportunities for generating fully automated quality systems with certain degree of autonomy. This definitely encourages transformation of traditional quality system to smart production system, especially when digitization process is recognized as one of the major strategic objectives.

Author details

Ercan Oztemel
Department of Industrial Engineering, Faculty of Engineering, Marmara University, Istanbul, Turkey

*Address all correspondence to: eoztemel@marmara.edu.tr

References

[1] Pham D.T., Oztemel E. Intelligent Quality Systems. London: Springer Verlag; 1996 London, ISBN: 3-540-76045-8

[2] Oztemel E., Tekez E. "multi-agent based integrated quality systems. In: 4th I*PROMS Virtual International Conference, 1-14 July 2008

[3] Oztemel E, Tekez Kurt E. A general framework of a reference model for intelligent integrated manufacturing systems (REMIMS). Engineering Applications of Artificial Intelligence. 2009;**22**(6):855-864. DOI: 10.1016/j.engappai.2008.10.013

[4] Rusenov B, Taneva A, Ganchev I, Petrov M. Machine Vision Systems for Intelligent Quality Control of Manufacturing Processes. In: International Scientific Conference on Engineering, Technologies and Systems (TECHSYS'17). 18 – 20 May 2017; Sofia. Bulgaria

[5] Z X, Yang D, Munro P. Knowledge-driven intelligent quality problem-solving system in the automotive industry. Advanced Engineering Informatics. 2018;**38**:441-457. DOI: 10.1016/j.aei.2018.08.013

[6] Bihi T, Luwes N, Kusakana K. Innovative Quality Management System for Flexible Manufacturing Systems. In: Open Innovations Conference (OI), 3-5 Oct. 2018, Johannesburg, South Africa. DOI: 10.1109/OI.2018.8535610.

[7] Tseng T, Aleti KR, Hu Z, Kwon Y. E-quality control: A support vector machines approach. Journal of Computational Design and Engineering. 2016;**3**:91-101

[8] Oztemel E. Intelligent manufacturing systems. In: Benyoucef L, Grabot B, editors. Artificial Intelligence Techniques for Networked Manufacturing Enterprises Management. London: Springer Verlag; 2010. pp. 1-41

[9] Shen WM, Norrie DH. Agent-based systems for intelligent manufacturing: A state-of-the-art survey. Knowledge and Information Systems. 1999;**1**(2):129-156

[10] Li B, Hou B, Yu W, Lu X, Yang C. Applications of artificial intelligence in intelligent manufacturing: A review. Frontiers of Information Technology and Electronics Engineering. 2017;**18**(1):86-96

[11] Oztemel E, Tekez E. Knowledge Protocols. In: Cunha M M, Olivera E F, Tavares A J, Ferreira L G, editors. Handbook of Research on Social Dimensions of Semantic Technologies and Web Services. IGI Global. 2009;304-324. ISBN: 978-1- 60566-650-1,

[12] Soon GK, On CK, Anthony P, Hamdan AR. A review on agent communication language. In: Alfred R., Lim Y., Ibrahim a., Anthony P. editors. Computational science and technology. Lecture notes in electrical engineering. 2019;481. Singapore: Springer Verlag. . DOI: 10.1007/978-981-13-2622-6_47

[13] Rowlands H, Packianather MS, Oztemel E. Using artificial neural networks for experimental design and off-line quality control. Journal of Systems Engineering. 1996;**6**:46-59

[14] Pham D T, Oztemel E.TEMPEX: An expert system for temperature control in an injection moulding process. Quality and Reliability Engineering International. 1992;8;1;9-15-DOI: 10.1002/qre.4680080104

[15] Wang J, Fu P, Gao RX. Machine vision intelligence for product defect inspection based on deep learning

and Hough transform. Journal of Manufacturing System. 2019;**51**:52-60. DOI: 10.1016/j.jmsy.2019.03.002

[16] Chesalin AN, Grodzenskiy SY, Nilovand MY. Tu P V. intelligent quality management tools for digital production and knowledge management system for their application. IOP Conference Series: Materials Science and Engineering. 2020;**862**. DOI: 10.1088/1757-899X/862/4/0420321

[17] Shen W, Norrie D H, Barthes J-P. Multi-Agent Systems for Concurrent Intelligent Design and Manufacturing. CRC Press; 2001. ISBN 9780748408825.

[18] Hossein IS, Hamid R. Intelligent QFD; a new approach for improving customer satisfaction. Lambert Academic Publishing. 2014 978-3659649356

[19] Bevilacqua. M, Ciarapica F F, Giacchetta G. A fuzzy-QFD approach to supplier selection. Journal of Purchasing and Supply Management. 2006;12;1;14-27

[20] Simić D, Svirčević V, Simić S. A hybrid evolutionary model for supplier assessment and selection in inbound logistics. Journal of Applied Logic. 2015;**13;2**:138-147

[21] Kuo. R J, Wang Y C, Tien F C. Integration of artificial neural network and MADA methods for green supplier selection. Journal of Cleaner Production. 2010;18;12;1161-1170 – DOI:10.1016/j.jclepro.2010.03.020

[22] Albers A, Gladysz B, Pinner T, Butenko V, Stürmlinger T. Procedure for defining the system of objectives in the initial phase of an industry 4.0 project focusing on intelligent quality control systems. Procedia CIRP. 2016;**52**:262-267. DOI: 10.1016/j.procir.2016.07.067

[23] Zabinski T, Mączka T, Kluska J, Madera M, Sęp J. Condition monitoring in industry 4.0 production systems - the idea of computational intelligence methods application. Procedia CIRP. 2019;**79**:63-67. DOI: 10.1016/j.procir.2019.02.012

[24] Yam RCM, Tse PW, Li L, Tu P. Intelligent predictive decision support system for condition-based maintenance. International Journal of Advanced Manufacturing Technology. 2001;**17**:383-391

[25] Nagata F, Tokuno K, Otsuka A, Ikeda T, Ochi H, Watanabe K, et al. Design tool of deep convolutional neural network for intelligent visual inspection. IOP Conference Series: Material Science and Engineering. 2018;**423**:012073

[26] Zhao T, Tiejun S, Xiao H, Linling S. Intelligent FMEA based on model FIORN. In: Annual Reliability and Maintainability Symposium (RAMS'04); 2004, Los Angeles; DOI:10.1109/RAMS.2004.1285480, pp:386-390

[27] Guimarees ACF, Ebecken NFF. FuzzyFTA: A fuzzy fault tree system for uncertainty analysis. Annals of Nuclear Energy. 1999;**26;6**:523-532. DOI: 10.1016/S0306-4549(98)00070-X

Chapter 2

Smart Manufacturing: Quality Control Perspectives

Joseph Evans Agolla

Abstract

Quality Control (QC) is a guideline or set of procedures intended to ensure that a manufactured product or performed service adheres to a defined set of quality criteria or meets the requirements of the client or customer. Smart manufacturing is where the work is interfaced work pieces and associated tools that include logistics operations, Cyber Physical Systems, Artificial Intelligence, and Big Data Analytic tools. These form the norm of manufacturing operations to generate large amounts of data, which are used for analysis and prediction. Therefore, help to optimise the quality of manufacturing operations and manufactured products. The change in technologies have, however, altered the traditional way of manufacturing process as well as QC systems. Therefore, to address the challenge of data reliability, the sensors, actuators and instruments used at various levels of integration in the manufacturing process often operating under adverse physical conditions need to provide adequate levels of data accuracy and precision. Methodologically, the Chapter followed critical literature review on QC concepts and Industry 4.0 revolution, thereby culminating into conceptual framework of QC in Smart Manufacturing, which is the main contribution of this Chapter.

Keywords: Industry 4.0, quality assurance, total quality management, organisation, artificial intelligence, big data analytics, logistics management, supply chains

1. Overview

The Chapter examines the Quality Control (QC) in Smart Manufacturing or Industry 4.0 Revolution. First, the Chapter begins with an examination of the historical development of QC from the Middle Ages (pre-Industry 1.0 revolution). Second, gives an overview of evolution of Smart Manufacturing and Quality Control, with emphasis on chronological trends of Industry revolutions up to date. Third, the concept of QC from the Middle Ages to 20th century. Fourth, it conceptualised the QC in Smart Manufacturing or Industry 4.0 revolution. Fifth, the Chapter gives an in-depth evolution of QC into Smart Quality Control Systems (SQCS) or Intelligent Quality Control Systems (IQCS), benefits, and lastly, summary of the Chapter.

1.1 Introduction

Historically Quality Control (QC) can be traced back to the times of pre-Industry 1.0 revolution when the modes of productions were still in their infant stage or Iron Age. During this period Human to Machine (H2M) interactions were still not common in the manufacturing/assembly lines. And humans were not specifically and strategically positioned within the production lines to ensure that products, which do not meet specifications are eliminated before end of the production processes. This is because the production was always manned by Artisans working with some few workers using simple and less mechanised tools of production. The manufacturing processes continued with the Artisans being solely responsible for the product quality, while the consumer was expected to apply the principle of '*Caveat emptor*' when buying products [1]. Although it should be noted that, there were some punitive measures put in place to guard against unscrupulous traders who could take advantage of the customers. This method of quality control continued until the advent of Industry 1.0 revolution, which brought some remarkable improvement on the ways and methods of productions. However, it is not possible to discuss QC without discussing how the modern quality systems have evolved. First, modern "Total Quality Management" emerged as a subset of Quality Control (QC), whose sole purpose was to ensure that entire production systems (from inputs to outputs) followed set standards. Total Quality Management (TQM) origin can be trailed to the early 1920s, the time statistical theory was introduced to product quality control. TQM's idea was further advanced in Japan in 1940s by three Americans namely Deming, Juran and Feigenbaum [2]. To date QC has followed the same standards of inspections of inputs (raw materials) before the production/manufacturing processes (assembly lines) and outputs (final products) reach the market.

The development of the present QC can be traced to the period of Hawthorne studies between 1924 and 1932, which highlighted the significance of social and psychological work climate [3]. In the same period, Shewhart also developed Statistical process control, which later became known as "Statistical Quality Process" (SQP). Statistical Quality Control (SQC) emphasised the products' design and production. Over the years though the concept of quality has developed into a discipline, a complex set of principles and assumed truths that define quality of goods and services is to be assessed, managed, delivered and assured [3]. During and until late into Industry 1.0 revolution, quality could be best described as "*caveat emptor*", which means, let "*the buyer be aware*". The manufacturers, artisans and industrialists produced goods of certain quality, but it was up to the consumer/buyer to appraise the quality of these goods. Thus, the consumer/buyer became responsible for the assurance of the goods they purchased [1, 2]. In the pre-industrial era, the quantity and quality of goods were the essential characteristics defining an economic transaction. In other words, "the qualities of the goods were known by their colour, sound, smell, taste, make, or shape [1, 4]. These forms of judgement made it problematic to differentiate the features that are the appropriate evaluators of those products. The problem is further compounded by the fact that people do differ very much; some person have clearer eyes, peculiar ears, noses, and tastes. In fact, the truth is every person having a good opinion of his/her faculties; therefore, it is difficult to find assessor to establish which is best" [4]. This approach makes quality to be more subjective and experiential. However, as progress made, through industrial evolution, and automation increased in the applications of manufactured goods, the level of product and process complexity, hence a new paradigm of quality control was borne, coinciding with a broader set of changes taking place under the realm of scientific management.

From Industry 1.0 to Industry 3.0 revolution, a lot of changes have been made in manufacturing lines/assemblies with the aim of producing products that meet customers specific needs. However, with the entry into Industry 4.0 revolution, organisations have moved from Human to Machines (H2M) production to Machine to Machine (M2M) intensive production, altering the way QC is managed in the manufacturing processes. Industry 4.0 revolution or Smart/digital manufacturing, with more emphasis on Big Data Analytic, Cyber-Physical Systems, 3-D printing, interface between M2M and Artificial Intelligence in the manufacturing processes [5]. Artificial intelligence (AI) can be conceived as the simulation of human intelligence in machines that are automated to think like humans and can imitate human behaviours [6]. This term may also refer to any machine, which displays attributes related to human minds such as learning and problem-solving.

1.2 Evolution of smart manufacturing and quality control

The fourth industrial revolution, or Industry 4.0 revolution (I4.0R), has become a reality today (**Figure 1**). The political debate about the term Industry 4.0 revolution focuses equally on the important and abstract objectives. For its promoters, Industry 4.0 revolution, though coined in Germany is not only about improving Germany's international competitiveness, but also perceived as means for solving some of the urgent global problems for example, climate change that has created new demand for the increased consumption of renewable and non-renewable resources. While some of the problems are specific national challenges such as, labour supply that is ever-changing due to demographic shifts [7, 8], Industry 4.0 revolution is focused on smart products, procedures, and processes (smart production). A key element of Industry 4.0 revolution is, therefore, the Smart Manufacturing (**Figure 1**). Smart Manufacturing or Industry 4.0 revolution are Cyber-Physical Systems, physical systems integrated with ICT components. These are autonomous machines that can make their own decisions based on machine learning algorithms and real-time data capture, analytics results, and recorded successfully past behaviours [9].

The Smart Manufacturing controls the fast-growing complexity, while also boosting production efficiency. Therefore, Smart Manufacturing is about direct communication between man, machine and resources to produce Smart products and services. Furthermore, Smart products know their manufacturing process and

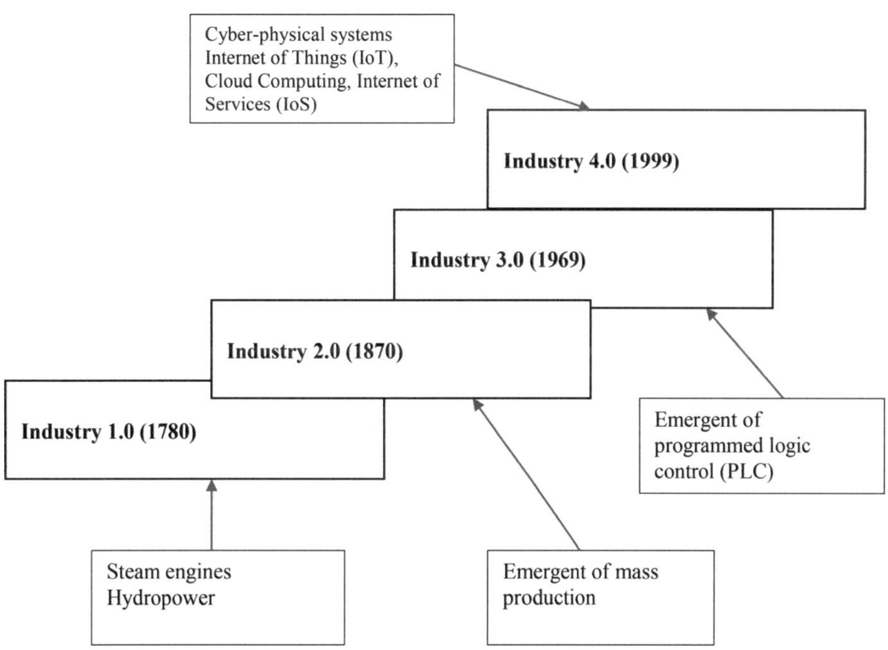

Figure 1.
Chronology of industry revolution. Source: Author's own illustration.

future applications. Equipped with these types of knowledge and intelligence, these gadgets actively support the production and documentation process. This will create value chain capable of answering questions such as (“*when was the product made, which parameters to be given to product, which destination is product intended for?*”) [10, 11, 13]. These interfaces to smart mobility, smart logistics, and smart grids make Smart Manufacturing an important element of future smart infrastructures. Conventional value chains will thereby be refined, and totally new business models will become established [12, 13].

Industry 4.0 revolution concept, therefore, encompasses not only value creation, but also work organisation, business models and downstream services. It performs this by using information technology networks of production, marketing and logistics. This enables it to capture all resources, production facilities and warehousing systems. The re-organisation thus, extends from the energy supply and smart power grids through to advanced mobility concepts (Smart mobility, Smart logistics) [12, 13]. However, on the technical side the concept is based on integrating Cyber-Physical Systems into production and logistics. In this Smart environment the concept of the Internet of Things (IoTs) and services that were already devised a decade ago have actually now become a reality. This process involves developing people and capital mobility, changing modes of production, consumption, learning, working and leisure, and increasing world-wide competition. In the subsequent subsection, we try to highlight the concept of quality control in Smart manufacturing.

2. Concept of quality control (middle age to 20th century)

Quality is as old as mankind on earth. It is possible that the quality of goods and services rendered has been monitored, either directly or indirectly since time immemorial [1, 2]. In the ancient Egyptians history, a commitment to quality in their pyramids was well demonstrated, similarly with the Greek architecture of the 5th century B.C. Such quality of work was evidenced in Roman-built cities, churches, bridges and roads that inspire the modern constructions [1, 2]. From the Middle Ages up to nineteenth century, the production of goods and services were predominantly confined to Artisans, a single person or small groups. These groups were mostly family-owned businesses, hence the responsibility for controlling the quality of a product or service rested with the Artisans or small groups [1, 2].

The quality of the goods and services rather followed the principles of “*caveat emptor*” as a single person controlled both the processes, leaving the buyer to determine the of the quality. This was pre-industrial revolution period. The phase comprising the time period up to about 1900, has been labelled the “*operator quality control*” period [14]. Then, the question is, what really motivated the Artisan or worker to continue producing quality goods and services? Perhaps, the worker took pride in total control of both number of products produced and the control of quality of such limited goods, hence a feeling of a sense of accomplishment, which lifted morale and motivated the worker to the new heights of excellence [2]. Therefore, controlling the quality of the products was embedded in the philosophy of the worker because a pride of workmanship was widespread. The “operator quality control” phase covered the entire pre-industry 1.0 revolution through to Industry 2.0 revolution.

However, beginning the early twentieth century through to 1930, a second wave evolved, that was referred toas the “*Foreman quality control*” [2, 14]. With this came Industrial revolution resulting in Mass production, which was based on the principle of the division of the labour specialisation. This principle placed emphasis on putting or assigning each worker according to their areas of skills and knowledge, for example, those workers who technicians, were grouped together, those skilled in the production were grouped also as such, and so on. That means no one person was entirely responsible for production processes, but rather a portion of it. But soon, the approach suffered some drawbacks as workers lost sense of accomplishment and pride in their work. This is because the workers could no longer control the entire production processes of the product being produced as before. Though most tasks were still not very complicated, and workers became skilled at the particular operations that they performed [2]. People were grouped together according to the tasks performed, for example, production units and assembly lines. A supervisor who directed the operation had the task of ensuring that quality was achieved. Foremen or supervisors controlled the quality of the product, and they were also responsible for operations in their span of control [2].

Then there came the period from 1920 to 1940 or World War II period, which saw the next phase in the evolution of QC. This period was known as the "Inspection quality control" [14]. The introduction of improved machines and equipment for industrial and manufacturing as a result of Industry 2.0 revolution, and increased demand for industrial and manufactured goods due to World War II, resulted in increased production volumes. However, as workers who were reporting to one foreman grew in numbers, it became apparent that these workers needed to be kept under close watch as a way to have control over the operations. This resulted in inspectors being assigned the tasks of quality check of the product after certain operations were completed [2]. Quality standards were set, and the inspectors compared the quality of the items produced against those standards. Any product found not meeting the set standards or, defective was put separately from those that met standard. The nonconforming products were reworked, if possible, or rejected altogether. It is at this period that the aspects of Statistical Quality Control (SQC) were being developed in the United States, however, it did not immediately gain wide usage in the United States industries [2]. Walter A. Shewhart of Bell Telephone Laboratories proposed the use of statistical charts to control the variables of a product, which later became to be known as "Control charts" (or Shewhart control charts). These played a vital role in Statistical Process Control. This was followed by H.F. Dodge and H.G. Romig, also from Bell Telephone Laboratories, who also pioneered work in the areas of acceptance sampling plans. These plans were to later to substitutes for 100 percent inspection [2].

The eve of 1930s saw the application of acceptance sampling plans in industry, both domestic and abroad. Walter Shewhart continued his efforts to promote to industry the fundamentals of statistical quality control (SQC). In 1929, he obtained the sponsorship of the American Society for Mechanical Engineers (ASME), the American Statistical Association (ASA), and the Institute of Mathematical Statistics (IMS) in creating the Joint Committee for the Development of Statistical Application in Engineering and Manufacturing. During these periods, the interest in the field of QC started to gain acceptance in England. The British Standards Institutions Standard 600 (BSIS-600) dealt with the applications of statistical methods to industrial standardisation and QC [2]. In the United States, J. Scanlon introduced the Scanlon plan, which dealt with improvement of the overall quality of work life [14]. Thereafter, the U.S. Food, Drug and Cosmetics Act 1938 had jurisdiction over procedures and practices in the areas of processing, manufacturing, and packaging.

The next phase of QC was the evolution process of the Statistical Quality Control (SQC), which took place between 1940 to 1960 [2, 14]. During these periods, the production of industrial and manufactured goods increased as a result of World War II and population explosion. However, because of mass production, 100 percent inspection became impossible, hence opening the way to the sampling plan [2].

In 1946, the American Society for Quality Control (ASQC) was established and immediately got renamed the American Society of Quality (ASQ). Then later on, the U.S. Military in 1950 developed a set of sampling inspection plans for attributes called MIL-STD-105A, which was modified to MIL-STD-105B, MIL-STD-105C, MIL-STD-105D and MIL-STD-105E. This was later followed by a set of sampling plans by the U.S. Military in 1957 [2].

After suffering humiliating defeat at the hands of the allied forces in the World War II in 1945, Japan wholeheartedly embraced the philosophy of SQC. This is after W. Edwards Deming visited Japan and lectured on these new ideas in 1950, which convinced Japanese engineers and top management of the importance of SQC as a means of gaining a competitive edge in the world market. The next person was J.M. Juran, another pioneer in QC who also visited Japan in 1954 and impressed upon Japanese on the strategic role top management plays in the achievement of a quality programme. The Japanese seized this opportunity and immediately realised the profound effects that these principles would have on the future of business, hence made a strong commitment to a massive programme of training and education [2].

The changes in quality swept through Industrial 1.0 to 2.0 revolutions, when the new paradigm that we all know as the '*quality control*' was borne. During this period, quality experts Edwards, Juran and Feigenbaum called upon the management to be more responsible, and responsive, to the issue of quality. [15] went further to state, "It is most important that top-management be quality minded". This sentiment was followed by [16], who echoed the significance of management commitment, "I submit that to enable QC to be really effective, we must work on making QC a member of the regular management team". Feigenbaum came up with an idea on organisation-wide efforts into the concept of "Total Quality Control" (TQC). The ideas behind TQC was that to provide genuine effectiveness, then real quality control management must start with the design of the product and end only when the product is finally with the consumer, who must remain satisfied, therefore quality was seen as everybody's business" [1, 17, 18]. However, it is worth noting that, the present QC manifested from Japanese, who revolutionise QC into the concept of "Total Quality Control". In their quest to revive home industries, after humiliating defeat in the hands Allied forces, the Japanese turned around their ailing industries after listening to Deming's lectures. Japanese TQC was manifestation of the third paradigm of the quality discipline, "Total Quality Management". This later on were embraced world over, though the concept originated in Japan.

2.1 Definition of quality control

The term 'quality' in essence means different things to different people. This is because people value different features of a product, some view quality as product package, price, colour, durability etc. However, [19] defines quality as, 'degree in which a set of inherent characteristics satisfies the requirements. The question is, what are these 'inherent characteristics' that a product must satisfy? The question can better be answered by; *first*, product values must meet or exceed the expectation of the final consumer. *Second*, overall quality- products, processes, systems, machinery and equipment must meet the statutory and contractual quality requirements [20]. Quality is not merely a monitoring tool by organisations, but rather should be viewed as a mechanism for anticipating problems, preventing them from occurring, and, ultimately, if they occur, solving them [20].

The importance of quality control ranges from good image, increase in sales volumes and competitiveness, good reputation, customer loyalty, just to mention a few. Global competition, customer heterogeneity, and technological change have altered the way the QC should be carried out in organisations. The traditional quality control has been rendered nearly obsolete by Industry 4.0 revolution technologies such as Big Data Analytics, Artificial Intelligent (AI), Cyber Physical Systems (CPS), Augmented Reality (AR), and Robo-Mate System gave birth to Smart Quality Control Systems (SQCS) or Intelligent Quality Control Systems (IQCS) as Manufacturing processes transforms itself to Smart Manufacturing [21, 22]. In Smart Manufacturing, AR can allow production managers to view productions KPIs and have an intra-factory overview of workstations and production lines in real-time for monitoring, identifying, analysing, diagnosing and resolving problems and flaws, a thing that used to be performed by human working in the production line. In the following subsection, we discuss the quality control in Smart Manufacturing.

2.2 Concept of quality control in smart manufacturing

Although the past three Industrial revolutions had quality control (QC) well entrenched in the manufacturing and industrial complexes, the application of the QC was more routine based on sampling plans and inspections. This was due to the fact that the three industrial revolutions' manufacturing and industry were characterised by mechanisation, waterpower, and steam power (Industry 1.0 revolution); Mass production, assembly line, and electricity (Industry 2.0 revolution). The distinctive features of these two Industrial revolutions were that they were both labour intensive, therefore, division of labour with emphasis on specialisation. However, Industry 3.0 revolution was anchored on computerisation and automation, hence eliminating some manual work that were carried out by human beings. Goods produced were of high quality compared to the previous two Industry revolutions as automation and computerisation were both introduced into the manufacturing and industrial complexes to aid in QC [23, 24]. The question is how do quality control (QC) works under Smart Manufacturing? In the subsequent section, the Chapter reviews some of the current literature to conceptualise QC in the Smart Manufacturing context.

2.3 Intelligent quality control systems

The concept of Intelligent Quality Control Systems (IQCS) or Smart quality control systems (SQCS) is founded on the premise that, in Smart Manufacturing production, quality control (QC), is driven by the infusion of Big Data Analytics, Artificial Intelligence (AI), Cyber-Physical Systems (CPS), Robotics and intensity of Human-to-Machine (H2M) interactions. The concept replaces the traditional QC systems in the manufacturing processes, as automation take over most of the operations or tasks that were routine tasks performed by human. Smart quality control is mainly executed to physically manage various Smart machines or tools through a cloud enabled platform. These technologies are capable of communicating both with the products (Smart products) and their environments. They are capable of detecting any slight defects and delays that could hamper manufacturing processes, and then communicate the same to the shopfloor, using fitted sensors [22, 25]. These gadgets work autonomously to create seamless communication between themselves. For example, [21, 26] installed sensors, utilised simulation and AI techniques assist in design and implementation of automatic machine model that predicts machine health status, which in turn can diagnoses any quality defects that could results from the machining failures. This result in a cost-effective solution in monitoring the production process to improve the quality of the products based on Industry 4.0 technologies.

Therefore, QC in Smart Manufacturing or Industry 4.0 revolution seem to take a different route as Industry 4.0 revolution is envisaged to leverage on a holistic automation, business information, and manufacturing execution architecture to improve industry with integration of all aspects of production and commerce across company boundaries for greater efficiency [27]. Industry 4.0 revolution is a complete departure from past three predecessors in several ways. First and foremost, Industry 4.0 revolution has come with Smart factories, Industrial Internet of Things, Smart Manufacturing, and Advanced Manufacturing, which were not experienced or witnessed in the past three successive Industrial revolutions. Second, Industry 4.0 revolution workplace emphasises so much on the Smart workers, Cyber Physical and Robotics in all the sphere of its Manufacturing and Industrial operations. The Internet of Things (Smart manufacturing, Additive Manufacturing, AI) have transformed the traditional production process of assembly lines with the introduction of asynchronous systems where predetermined workflows based on production work orders are running enterprise business systems [27]. Hence, making production steps that are centrally in communication to each Manufacturing station, which is harmonised with the assembly line.

In contrast, asynchronous manufacturing is based on I4.0 revolution concept in which components in the production flow using auto-identification technology to inform each machine and operator on what needs to be done to produce customised end product. This activity takes place at each step of the production process. In this process, the machines are more flexible, which make them adaptable to the requirements for the part being made at each production steps. This entire concept is a product of Industry 4.0 revolution. The systems assist in achieving a highly flexible, lean, and agile production process that allow for a variety of distinctive products to be produced in the same production facility. The process is based on the premise of profitable mass customisation that enables the production of small lots (even as small as single unique item). This is due to the ability to rapidly configure machines to adapt to customer-supplied specifications and additive manufacturing [27]. **Figure 2** below gives a snapshot of Manufacturing production process and the quality control under Smart Manufacturing. Inputs- denotes Smart raw materials, and Smart workers that are capable to communicating with Robotics to execute the tasks. Such systems comprise production facilities, storage systems and smart machines which trigger actions, exchange information complete autonomously and are able to control each other independently [8, 28].

Smart raw materials will be detected by machines without necessarily having to be verified or inspected as the case in the past. The machines fitted with sensors will be able to differentiate between quality inputs (Smart raw materials) and defects, if possible reworked, or discarded all together, a thing that was formerly done by human beings in the traditional manufacturing set up (see, **Figure 2**). Inputs will have Smart workers, who are capable of interacting with computers and Robots. The Smart Manufacturing is fully equipped with actors, sensors and CPS where

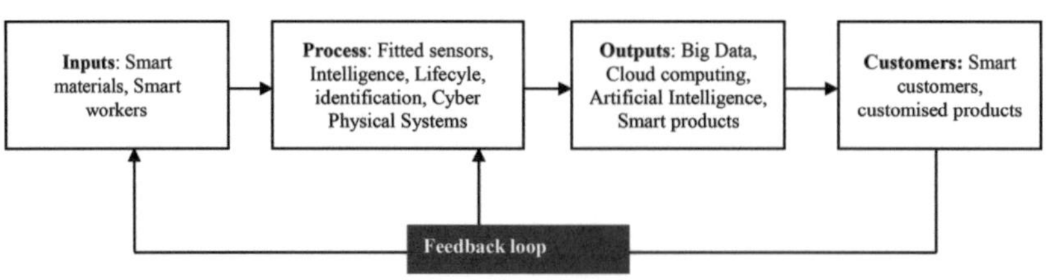

Figure 2.
Intelligent quality control Systems in Smart Manufacturing. Source: Author's own illustration.

"human beings, machines and resources communicate with each other as naturally as in a social network" [8, 28] as shown in **Figure 2**.

In **Figure 2**, Smart Manufacturing process begins with the input as smart material (because these materials are fitted with microchips, sensors), which enable them to be recognised and detected by the intelligent machines. The fact is, the material can be configured or reconfigured according to the Smart Manufacturing requirements, if found not to meet the specific product manufacturing specifications, then it can be discarded or reworked. This allows for the smooth flow of manufacturing process. This results in an improved finished product quality and reduced level of production errors [5, 25]. The implementation in the technology production process namely, ICTs, sensors technology and robotic technology, have the ability to record the production process in each element (instead of sampling and control) and detecting errors that occur during the process [25]. If errors occur or are detected, the machines can be adjusted in real time accordingly.

In the manufacturing process, there are Smart machining, Smart monitoring, Smart control, and Scheduling (**Figure 2**). Cyber-Physical Systems enable Smart machine tools to capture the real-time data and send it to a cloud-based central system so that machine tools and their twined services could be synchronised to provide to Smart Manufacturing solutions. While, Smart monitoring, monitors the operations, maintenance, and optimal scheduling of manufacturing systems. Smart monitoring assist in Smart Manufacturing by giving warnings/alerts if some abnormality occurs to machines/tools. In addition, Smart control, though can be executed to physically manage various smart machines or robot through Cloud enabled platform [5] but do allow the end-users to switch off a machine or robot via their Smartphones [29]. This allows the decisions to be reflected in frontline manufacturing sites such as robot-based assembly lines or Smart machines (**Figure 2**). Then finally, Smart scheduling which includes advanced models and algorithms draw on data captured from sensors [5]. These data-driven techniques and advanced decision architecture is used in smart scheduling. **Figure 2**, with the assistance of data input mechanisms, the output resolutions are fed back to the parties through various means (feed loop) [5, 30]. **Figure 2** for example, comprises Big Data, Clouding Computing, Internet, Simulation, Artificial Intelligence, and System Integration, which represent technologies, such as Additive Manufacturing, Autonomous Machines, and Human -to-Man (H2M) integration. These produce faster, stronger and more consistent than workers with a combination of new sensors and actuators and extensive data analysis [25].

In **Figure 2** above, process represents transformation process of Smart raw materials into final products. Industry 4.0 revolution comprises a high-resolution, adaptive production control (APC) such as Smart Control that can be achieved through development of Cyber-Physical production control systems [10]. In addition, Smart control is mainly executed to physically manage various Smart machines or tools through a cloud enabled platform [31]. End-users (Smart customers are able to get a smart product, which are tailored-made according to their personalised needs. In addition, smart customers are able to interact with smart products from smart manufacturing and could easily identify with such products. And if the product fails to meet their specifications [29], the decisions could then be timely reflected in frontline manufacturing sites such as robot-based assembly lines or smart machines [32], as shown in **Figure 2** above. For instance, the Smart quality control in Smart Manufacturing is well illustrated by Changying Precision Technology Company's factory in Dongguan city. This is the first unmanned factory run by computer-controlled robots, numerical control machining equipment, unmanned transport trucks, and automated warehouse equipment. It is said that about six hundred human assembly-line workers were replaced with this automation alone. The result was a fivefold reduction in manufacturing errors and an increase in production of more than 250 percent [27, 33]. This is a typical example of how quality control (QC) in Smart Manufacturing has been operationalised.

Smart Manufacturing or Industry 4.0 revolution is built around the concept of self-control or managing production processes requires open software and communications standards that allow sensors, controllers, people, machines, equipment, logistics systems, and products to communicate and cooperate with each other directly [5, 27]. This simply means the use of human beings in the manufacturing process particularly in the production/manufacturing processes as quality control inspectors, is minimised, if not eliminated. However, to embrace Smart Manufacturing in sustainable way, require that Manufacturing industries adopt technologies transformations with training and development programmes in order to fit their workforce with the new workplace requirements, such as new tools and technologies [11, 34]. This will ensure that gaps in skills and knowledge created by the Smart Manufacturing technologies do not have serious impacts on the workforce work life. Therefore, the implementation of Smart Quality Control Systems (SQCS) or Intelligent Quality Control Systems (IQCS) in Industry 4.0 requires further employee skills and competencies, such as ICT know-how, interdisciplinary competencies and special personality traits [34, 35]. This is because Human-to-Machine (H2M) collaboration that is necessary as some production tasks are too unstructured to be fully automatised.

In the production assembly/manufacturing assembly, Virtual Reality (VR) and operator create 'cognitive interaction'. For example, VR technology provides a combination of interactive reality and advanced simulations that can replicate a design, assembly, or manufacturing environment and allow the smart operator to interact with any (Machine tools, production line, hand-tool, a robot, a factory), with reduced risk and real time feedback as shown in **Figure 2** [11]. In addition, VR, at product assembly stage, CAD models of parts, hand tools, and assemblies can be transformed into interactive simulations (assembly sequence). This can be used in the training of operators working in a complex assembly tasks, and at product manufacturing stage. [11] opine that VR brings to life the "virtual factory" as an integrated simulation model of the major subsystems of a factory layout (such as arrangements of machinery, equipment and inventories for smooth flow of work, material and finished products). These arrangements form continuous communication between humans to machines and products during the production process. This is enabled by Cyber-Physical Production Systems (CPPS) in order to execute its tasks. The overall aim is to decrease cost, time efficiency, and improve product quality, which requires a broad understanding of the enabling technologies as well as methods and tools [13].

Products in Smart Manufacturing are 'Smart', with embedded sensorics that is used via wireless network for real-time data collection for localisation, for measuring product state and environment conditions [9]. In **Figure 2**, Smart products have control and processing capabilities, thus control their logistical path through the production and even optimise the production workflow. In addition, Smart products are capable of monitoring their own state during the whole lifecycle, including their lifetime or application [9].

Already in use is the intelligent Quality Control Systems (IQCS), which has replaced the traditional QC in the manufacturing processes. In Smart Manufacturing, all aspects pertaining to products quality control are first defined. In the first step, the technical requirements for the quality control system in development are defined and documented. In addition, the final document is intended for as a working-document supporting the requirements process during the development of the quality control systems [13]. The introduction of intelligent-based quality control necessitates integration steps within and outside the manufacturing industry. It affects sensors and actuators as well as general manufacturing processes, like information and documentation flows. Furthermore, manufacturing partners or customers have to be integrated into the development as they are all part of the overall value chain (see **Figure 2**).

2.4 Benefits of intelligent quality control systems

IQCS brings with it several benefits to those organisations that will be able to adapt the new technologies in Smart Manufacturing processes compared to traditional quality control has been part of the manufacturing processes in the previous three industrial revolutions. Such benefits are summarised as follows:

- "Time to market" to develop, produce and market new products and services, requiring higher and faster innovation capability [36]. This is due to the cutting-edge technologies such Additive manufacturing, Industrial Internet of Things (IIoTs), Augmented Reality and Virtual Reality. These have eliminated wastes that were formerly associated with human errors hence creating lean production of products that are competitive globally. Augmented Reality (AR) assist in reducing defects, rework and redundant inspection by offering intuitive information and combining operator intelligence and flexibility with error-proofing systems to increase efficiency of manual work steps, while improving the quality of work [9, 11].

- Increased "customisation" to satisfy individual consumer demands, in a buyer' market, not anymore a seller's one, leading to higher product individualisation; meaning products may not need to be produced in mass as before, because the manufacturers will be able to produce very small series (single product if needed). This technology provides fast configuration of machines and production process, as well as their adaptation to customer requirement [25].

- Higher "flexibility" with faster and more versatile production processes able to produce smaller lot quantities with high quality and a cost-effective way [9, 22, 36].

- "Decentralised" decision making with fewer organisational hierarchies be reduced.

- Increased resource "efficiency" by using more efficient and closed loops, regenerative, and restorative physical and economic cycles, where products and raw materials retain their physical characteristics and value as much as possible [9, 36].

3. Summary

Quality control (QC) has evolved from the Middle Ages to the present time, with the changes in manufacturing industries. As industrial revolutions transformed itself, so is the QC systems. First QC started as '*caveat emptor*', whereby the control was entirely in the hands of Artisan, and it was the responsibility of the customer to ensure that the product is of quality. This was followed by punitive measures imposed on the Artisan who produces inferior products. From there came the '*operator quality control*' (OQC). The operator was to ensure that a product meet certain standards of the quality. Due to the failure of the two and the emergence of industrial 1.0 revolution, another QC principle was adopted, '*foreman quality control*' (FQC) fill the void that could not be filled by the two concepts. On the onset of Industry 2.0 revolution, there emerged '*Statistical Quality Process*', which later turned to be known as '*Statistical Quality Control*' (SQC). The discovery of electricity paved ways for complex machineries and industrial complex, hence the need to use data in the manufacturing processes. Then from middle of Industry 2.0 revolution to the end of Industry 3.0 revolution, QC transformed itself from just merely a control tool, to '*Total Quality Management*' (TQM) with emphasis on the entire manufacturing processes.

However, in contrast to the previous Industrial revolutions, Industry 4.0 revolution or Smart Manufacturing has completely revolutionised how QC in manufacturing processes is practiced. Traditional QC systems has now given birth to Smart Quality Control Systems (SQCS) or Intelligent Quality Control Systems (IQCS), where machines have taken over most of the roles performed by human in the manufacturing. Technological development has made production/manufacturing processes to be more complex and complicated, yet simple in terms of networked processes created by the application of Cyber-Physical Systems, Additive Manufacturing, Artificial Intelligent, Augmented Reality, and Virtual Reality. These networked technologies assume the tasks and roles that were manually performed in the manufacturing processes, thereby eliminating human errors that were common in the products design and development. The use of 3-D or 4-D printing now enable manufacturers to produce prototypes and proof of concept designs, which simplify and speed up the processes of new product design and manufacturing [24]. Hence, resulting in the following benefits to the organisations; lower production costs, low logistic costs, and quality management costs, and others are improved customer responsiveness, customisation of mass production without significantly increasing production costs, more efficient use of natural resources and energy, and more friendlier working environment. These are some of the benefits of IQCS as Smart Manufacturing.

Author details

Joseph Evans Agolla
Department of Management, School of Business and Management Studies, Botswana Open University, Botswana

*Address all correspondence to: nyagonya2009@gmail.com

References

[1] Dooley, K. (2000), The paradigms of quality: Evolution and revolution in the history of the discipline. *Advances in the Management of Organisation Quality*, Vol.5 JAI Press:1-28.

[2] Mitra, A. (2016), *Fundamentals of Quality Control and Improvement*. 4Th ed. John Wiley & Sons. www.wiley.com/go/mitra/QualityControl4e. Accessed 29 September 2020. ISBN:978-1-118-70514-8.

[3] Martínez-Lorente, A.R., Dewhurst, F. and Dale, B.G. (1998), Total quality management: origins and evolution of the term. *The TQM Magazine*, Vol. 10 No. 5, pp. 378-386. https://doi.org/10.1108/09544789810231261.

[4] Barbon, N. (1690), A discourse of trade. In Dooley, K. (2000), "The paradigms of quality: Evolution and revolution in the history of the discipline." *Advances in the Management of Organisation Quality*, Vol.5 JAI Press:1-28.

[5] Zheng, P., Wang *et al*. (2018), Smart manufacturing systems for Industry 4.0: a conceptual framework, scenarios and future perspectives. Front. Mech. Eng. DOI 10.1007/s11465-000-0000-0. Accessed 19th August 2020.

[6] Heng, S. (2020), Industry 4.0 creating a buzz in western hemisphere: But watch out for China pulling into the fast lane. In: Machado, Carolina and Davim, J. Paulo (Eds.). "Industry 4.0: Challenges, Trends and Solutions in Management and Engineering". Taylor & Francis. Boca Raton. pp 43-76.

[7] Heng, S. (2014), Deutsche Bank AG, Deutsche Bank Research, 60262 Frankfurt am Main, Germany. All rights reserved. When quoting please cite "Deutsche Bank Research". Print: ISSN 1612-314X / Internet/E-mail: ISSN 1612-3158. Accessed 10th July 2020.

[8] Kagermann, H., Wahlster, W., Helbig, J. (2013), Recommendations for implementing the strategic initiative INDUSTRIE 4.0 - Final report of the Industrie 4.0 Working Group.

[9] Rojko, A. (2017), Industry 4.0 concept: Background and overview. Special focus paper. *International Journal of Innovation Management*, Vol.11 No.5, p.77-90.

[10] Stich, V., Hering, N. and Meißner, J. (2015), Cyber physical production control: Transparency and high resolution in production control. in IFIP *Advances in Information and Communication Technology* Vol. 459: pp. 308-315.

[11] Romero, D., Stahre, J., Wuest, T., Noran et al. (2016), Towards an Ooperator 4.0 typology: A Human-centric perspective on the fourth industrial revolution technologies. CIE46 Proceedings, 29-31 October 2016, Tianjin / China, ISSN 2164-8670 CD-ROM, ISSN 2164-8689 on-line.

[12] Kang, H.S., Lee, J.Y., Choi, S. et al. (2016), Smart manufacturing: Past research, present findings, and future directions. *International Journal of Precision Engineering and Manufacturing-Green Technology*, Vol. 3, p.111-128. https://doi.org/10.1007/s40684-016-0015-5. Accessed 10 November 2020.

[13] Albers, A., Gladysz, B., Pinner, T., Butenko, V. and Stürmlinger, T. (2016), Procedure for defining the system of objectives in the initial phase of an Industry 4.0 project focusing on intelligent quality control systems. Procedia CIRP 52:262-267. Peer-review under responsibility of the scientific committee of the Changeable, Agile, Reconfigurable & Virtual Production Conference 2016. ELSEVIER. www.elsevier.com. Accessed 29 September 2020.

[14] Feigenbaum, A.V. (1983), *Total Quality Control*. 3rd ed., McGraw-Hill, New York.

[15] Juran, J. (1945), *Management of inspection and quality control*. NY: Harper & Bros.

[16] Feigenbaum, A. (1952). The significance of quality control to present day management. Industrial Quality Control, Vol.8 No.6, pp. 29-34.

[17] Feigenbaum, A. (1957), The Challenge of total quality control. *Industrial Quality Control*, Vol. 13 No.11, pp. 47-50.

[18] Feigenbaum, A. (1961), *Total quality control*.1st Ed. NY: McGraw-Hill.

[19] International Standards Organisation 9001 (1994), Quality systems - Model for quality assurance in design, development, production, installation and servicing. https://www.iso.org/standard/16534.html. Accessed 21 August 2020.

[20] Godina, R. and Matias, J.C.O. (2019), Quality Control in the Context of Industry 4.0: https://www.researchgate.net/publication/332455424. Accessed 6 October 2020. DOI: 10.1007/978-3-030-14973-4_17.

[21] Sader, S., Husti, I. and Daroczi, M. (2019), Quality Management Practices in the era of Industry4.0 revolution. Zeszyty Naukowe Politechniki Częstochowskiej Zarządzanie Nr 35 (2019) s. 117-126 dostępne na: http://www.wz.pcz.pl/znwz. Accessed 12th September 2020.

[22] Schmidt, R., *et al*. (2015), Industry 4.0 – Potential for Creating Smart Products: Empirical Research Results.www.researchgate.net/publications/274894802. Accessed on 6th October 2020.

[23] Schwab, K. (2015), The Fourth Industrial Revolution: What it means and how to respond. www.foreignaffairs.com/articles/2015-12-12/fourth-industrial-relation. Accessed 30th September 2020.

[24] Gilchrist, A. (2016), Industry 4.0: The Industrial Internet of Things, Springer, Heidelberg.

[25] Karabegovic, I. (2018), The role of Industrial and Service Robots in the 4th Industrial Revolution- "Industry 4.0". ACTA TECHNICA CORVINIENSIS – Bulletin of Engineering Tome XI. Fascicule 2.

[26] Kuo, C.J., Ting, K.C. and Chen H.M. (2017), Automatic Machine Status Prediction in the Era of Industry 4.0: Case Study of Machines in a Spring Factory, "Journal of Systems Architecture", Vol. 81. DOI: 10.1016/j.sysarc.2017.10.007. Accessed 5th October 2020.

[27] Lydon, B. (2016), Industry 4: Intelligent and flexible production: digitisation improves manufacturing responsiveness, quality, and efficiency. https://www.isa.org/intech-home/2016/may-june/features/industry-4-0-intelligent-and-flexible-production. Accessed on 29 September 2020.

[28] Foidl, H. and Felderer, M. (n.d), Research challenges of Industry 4.0 for quality management. https://d1wqtxts1xzle7.cloudfront.net/49809226. Accessed 5th October 2020.

[29] Wang, L.H. (2013), "Machine availability monitoring and machining process planning towards Cloud manufacturing." *CIRP Journal of Manufacturing Science and Technology*, Vol. 6, pp. 263-273.

[30] Xu, M., David, J.M., and Kim, S.H. (2018), The Fourth Industrial Revolution: Opportunities and

Challenges. *International Journal of Financial Research* Vol. 9, No. 2, 90-95.

[31] Makarov, O., Langmann, R., Nesterenko, S. and Frank, B. (2014), Problems of the time deterministic in applications for process control from the cloud, *International Journal of Online Engineering*, Vol. 10, pp. 70-73.

[32] Wu, D.Z., Rosen, D.W., Wang, L.H. and Schaefer, D. (2015), "Cloud-based design and manufacturing: A new paradigm in digital manufacturing and design innovation." *Computer-Aided Design*, Vol. 59, pp. 1-14, 2015.

[33] Lorenz, M., Rüßmann, M., Strack, R., Lueth, K.L. and Bolle, M. (2015), Man and Machine in Industry 4.0: How will technology transform the industrial workforce through 2025. Boston Consulting Group, Inc. https://image-src.bcg.com/Images/BCG_Man_and_Machine_in_Industry_4_0_Sep_2015_tcm9-61676.pdf. Accessed 3rd October 2020.

[34] Gunasekaran, A., Subramanian, N. and Ngai, W.T.E. (2019), Quality management in the 21st century enterprises: research pathway towards Industry 4.0. *International Journal of Production Economics*, 207. pp. 125-129. ISSN 0925-5273.

[35] Veile, J.W., Kiel, D., Müller, J.M.and Voigt, K-I. (2019), Lessons learned from Industry 4.0 implementation in the German manufacturing industry. *Journal of Manufacturing Technology Management*. DOI: 10.1108/JMTM-08-2018-0270.

[36] Fonseca, L.M. (2018), Industry 4.0 and the digital society: concepts, dimensions and envisioned benefits: Accessed 5th October 2020.DOI: 10.2478/picbe-2018-0034.

Chapter 3

An Intelligent Access Control Model

Shadha Mohamed Sulaiyam ALAmri

Abstract

Cybersecurity is a critical issue as the world is moving towered IR4 era (Industrial Revaluation 4.0) where technology is involved, and access to the internet is an imperative need. The traditional computing systems are not able to meet the huge computing demand and growing data (Big-Data). Therefore; new technologies have been evolved such as cloud computing. This chapter is exploring the need for a dynamic access control approach to enhance the Cybersecurity. The scope in this chapter is focusing on IaaS (Infrastructure as a Service) layer of cloud computing. The research approach aims to enhance the basic ABAC (Attribute-Based Access Control) model by adding a context-aware feature and SoD principle. The enhanced model called **ABACsh**. This proposed enhancement is implemented through a framework based on AI (Artificial Intelligent) to meet the requirements of dynamic systems. The framework is tested in the OpenStack testbed. The results show better performance in the term of computation speed.

Keywords: Cybersecurity, AI, ABAC, formal logic, IaaS, cloud computing, OpenStack

1. Introduction

Industrial revolution 4 (IR 4.0) utilizes technology in different aspects. As per the world economic forum, three principle technology drivers in the industrial production: connectivity, intelligence and flexible automation where big data is one of the value drivers and IoT is one of the scale-ups enablers [1]. To summarize one scenario of the embedded technology in the industry is the implementation of IoT systems. Internet of Things (IoT) starts with a collection of sensors used to collect information from the surrounded environment. For example, a temperature sensor used to collect the atmosphere temperature during the day by taking three reads for six months for the purpose of studying climate change. The collected data will be sent to central storage such as cloud computing technology to get the advantage of accessing the data anywhere and anytime. There is a need for a network connective that allows distributed components to be connected. Mostly the collected data is a type of big-date as they might collect temperature reading from a different site in the globe and for a long time might be years. That big-data requires some analysis where the traditional analytical system might not manage to absorb its huge records, therefore; there is a need to utilize the features of artificial intelligence filed in data-science. This example shows how several technologies are used in order to be used in analyzing the big-data collected from different sites.

As there are distributed systems and the internet connection is used, cybersecurity becomes a critical aspect, especially when there are some economic benefits. There are many security principles which might be tackled in order to enhance the cybersecurity of the systems, however, access control is one of the major aspects as there will be a need to restrict the access to the system as there is a distributed environment when it comes to IoT deployment. One of the optimal access control models to be used in this case is attribute-based access control model (ABAC) [1, 2].

This chapter introduces intelligent attribute-based access control model tested in the cloud computing environment. Section 2 discusses the introduced enhancements to the basic ABAC model. Section 3 illustrates how inelegant is introduced in the proposed ABAC. Finally, an empirical experiment is demonstrated in Section 4 where OpenStack (cloud environment) is used to discuss the efficiency of the proposed approach.

2. The proposed enhancement in ABAC model

2.1 Context-aware analytical study

Context-aware system has verity of definitions based on the study scope. In access control field, context-aware allowing a dynamic permission to access an object based on some attributes related to the user context [3]. The context can be extracted from the system environment by using 5W1H (who, where, what, why, when and how) [4]. The context attributes are a finite set which reflect the system and differs from the attributes related to the subject and the object as per the researcher in [5]. However, other authors consider the rule enforcement through Attribute-Based Access Control (ABAC) is based on the attributes of both the subject and the object [6]. Therefore; this section is investigating context-aware concept in access control.

An ubiquitous application with RBAC extension has been investigated by Kim's [3] where state checking matrix is used to build a context-aware agent. Two cases are defined to deploy context-awareness. The first one is through giving privilege up-on the user context, such as location and time. The second one is changing resource permissions up-on the system information, such as network bandwidth and memory usage. Another work proposed by Kim in this filed, called CIAAC (Context Information-based Application Access Control) [4]. CIAAC designed to separate processing logic and business from context awareness and access control policy. CIAAC add flexibility to business application which support dynamic access control policy. This feature allows to satisfy the demand of external security environment. However, the potential drawbacks of CIAAC have not yet been evaluated. Another technique was proposed by Li in his thesis [7] to meet the scope of mobile cloud environment based on Attribute-Based Encryption (ABE). Li defines context-aware terminology to cover the user context-information in addition to the environment such as location and time.

As per the literature, encryption techniques such as ABE introduce several limitations which effect the overall system efficiency such as the overhead caused by bilinear pairing due to its heavy computation [8]. In addition to that ABE cannot attain fine-grained control [9]. Another related work done by AL Kukhun [10] considering pervasive systems where XACML language is used to build a model to extend RBAC that can facilitate context-aware features. However, RBAC extinctions approaches do not satisfy usability, situation awareness, and improving access opportunities. It can be observed that location and time are used as context-aware parameters in most related work on context-aware access control models. Liu and

Wang [11] present the Fine-grained Context-aware Access (FCAC) model for Health Care and Life Sciences (HCLS) using specific communication technology based on linked data. FCAC is based on two main components: an ontology base, and access policy with XACML.

It is observed from the state of art that context attributes are linked to the system environment rather than subject-attributes or object-attributes. Venkatasubramanian et al. [12] investigate context-aware to distinguish between the traditional authorization models and their proposed criticality-aware as they take into consideration the context of the whole system. Their criticality-aware (CAAC) is based on RBAC concepts. Choi [13] used access-aware in cloud computing. Choi recommends an ontology-based Access Control Model (onto-ACM). Compared to C-RBAC (Context-aware RBAC), onto-ACM can grant the role inheritance by administrator and user, whereas C-RBAC grants the role by administrator only.

2.2 The proposed context-aware deployment in $\mathbf{ABAC_{sh}}$

As per the related work investigated in Section 2.1, we can conclude that to deploy an efficient context-ware feature, the attributes should be related to the system environment. Context-aware will add a flexibility to dynamic systems where the users and privileges keep changing such as the case in IaaS. ABAC is the basic access control type which can support the context-awareness. Therefore, we are not recommending RBAX extensions.

The proposed $\mathbf{ABAC_{sh}}$ model is adding context-aware through two phases. The first phase defines the context-attribute set. Each context-attribute consists of an attribute name and an attribute value. The context attributes-names set is predefined by the system administrator based on system critical information and characteristics. Context-attributes differ from the environment attributes in that the latter values are predefined by the administrator, whereas context-attribute values are updated based on the system states, where an embedded sensor captures the context information. For example, for the context-aware attribute named memory, its value will be updated based on the system memory measurements. The context attribute can reflect CPU clock, desk space, network zone, or data and time. In the second phase, context-awareness will be defined as one of the configuration points in the proposed $\mathbf{ABAC_{sh}}$ system to enforce the use of context in the access-control decision.

2.3 Critical analysis of SoD in ABAC

In an environment that allows policy combination, a user is authorized to act in more than one role or trigger more than one operation simultaneously. Policy combination might lead to policy conflict, as some actions violate the overall policy if they are committed at the same time. Therefore, constraints should be configured to manage this possibility.

The Separation of Duty (SoD) principle is used in such scenarios to prevent misuse of the system by limiting the user to the least privilege necessary to perform their required tasks. The least privilege principle limits the access of the subject during an operation on a specific task to be within the minimum resources, lowest privileges, and specified period of time. Several security enhancements can be gained from SoD, such as fraud prevention and error minimization [14–16].

There are two main types of SoD: static, and dynamic. Static-SoD (SSoD) will list the conflicting roles which cannot be executed by the same user at the same time, whereas dynamic-SoD (DSoD) enforces the control at the time of access-request. In an RBAC model, roles and role relations are defined in advance during the policy engineering process. For SSoD in RBAC, SSoD relations place constraints on the user-to-rule assignment function, where one user can be assigned a specific set of roles and be excluded from another set of roles. Otherwise, two or more users are required to be involved in accomplishing sensitive tasks, since it is less likely that multiple parties will issue a fraud attack. In the DSoD relation, the capabilities for one user are restricted to being activated during a specific user session, i.e. the same user cannot perform two roles simultaneously [17, 18].

Although in RBAC, SSoD and DSoD relations offer some advancement in control over identity-based systems, security issues remain. The most accommodating form of SoD is History based (HSoD). Although, enforcing it in a static based access control management environment such as RBAC is di_cult, if not impossible [19, 20]. One role of a HSoD is that it prevents an object from being accessed by the same subject a certain number of times [21]. Therefore, we assume that the ABAC model concept has the characteristics of supporting certain types of SoD. The following section will investigate efforts in the literature to involve the SoD principle in ABAC.

A significant amount of research has been conducted regarding the principle of Separation of Duty (SoD) in RBAC; however, SoD deployment in ABAC remains a problem [22]. One of the earliest related works in specifying constraints in ABAC is illustrated through ABAC configuration-points [5]. Nevertheless, their proposed constraint settings are event-specific during attribute assignment and/or modification of the object and subject. This method is similar to the RBAC constraints-setting concept, where the allowed roles are activated for a specific user session after the roles are assigned to the users.

The author of ABCL proposed an event-independent constraint language based on conflicting relations of attribute values such as mutual exclusion and precondition [23]. ABCL language specifies restrictions either on a single set of attribute values or on a set of values of different attributes within the same entity. The usefulness of ABCL language has been validated through case studies. However, it lacks a framework or a formal model that illustrates its implementation. Dynamic Separation of Duties (DSoD) is more appropriate to cloud computing, and it also meets the dynamic nature of ABAC. Nguyen [24] has carried out interesting research on DSoD and proposed DSoD deployment through Provenance based Access Control (PBAC). His work is basically proposing a means to capture and utilize the information needed in the SoD enforcement, as previous work in the area assumes that the information is ready without demonstrating how to prepare it. Some of the previous work related to dynamic-based SoD is ObjDSoD, which is based on the object, and where the enforcement is constructed on a set consisting of conflicting-roles and a conicting-action on these roles. Therefore, the subject will not be allowed to perform an action on an object if that action is in the set of action role conflict. Another approach is OpsDSoD based on operations. This is a task-aware that involves an action-role conflict set, thus it differs from ObjDSoD by limiting the user to perform the needed actions for a particle task even though they have more privileges. A third approach is HDSoD, which combines ObjDSoD and OpsDSoD. Further, HDSoD is object-aware and a task aware. HDSoD is order-aware, where order-dependency conflict is triggered if the order is essential for a sequence of sub-tasks. Nguyen in [24] extended HDSoD by adding dependence-path-aware and past attribute-aware in their DSoD which is used in Provenance-based Access Control (PBAC).

Event pattern and response relations called obligations are introduced by Ferraiolo, Atluri, and Gavrila in their policy machine research [19], which can enforce some forms of HSoD in their access control framework. Obligations have a set of conditions that are specified by the event pattern under which the state of the policy is obligated to change; only if this set matches the surrounding context, the operation on an object can be executed. There are two recognized standards can be applied to the ABAC concept: Extensible Access Control Markup Language (XACML), and Next Generation Access Control (NGAC) [25]. XACML does not show any support for DSoD constraint, while NGAC does show some support to DSoD through a Prohibitions (Denies)-relation, which includes a set of denying relations that specify privilege exceptions where a user that is allowed to run capability (x) will be prohibited from running capability (y).

2.4 SoD design and deployment in access control system

It is most likely that the formulation of SoD requirements are prepared by the administrator based on the business rules. An example of such a rule is, person may not approve his or her own purchase order [26]. SoD deployment can be involved with different layers of an access control system. It can be designed within administrative-level policies and procedures, or it can be used within logical or technical mechanism access-control restriction points [15].

Based on recommendations regarding SoD implementation to traverse its limitation in RBAC [20], several techniques have been explored, such as grouping concept, membership control, activation control, history control, and labels. However, in ABAC, the grouping concept will not be appropriate as grouping restricts an attributes flexible nature. Membership control cannot be adopted by ABAC as it is not role-centric. Though, the activation control concept has been adopted into SoD specifications in ABAC by Jin [5] and Bijon [23]. Ferraiolo et al. [19] describe a relation between entities that can be used in History based SoD deployment. Whereas Biswas et al. [27] point out that label concept can be used to enforce SoD in their proposed label-based access control in an ABAC. There are several obstacles in designing and implementing SoD, as it is an application-oriented policy where the business rules indicate the critical tasks which require SoD enforcement. Another challenge is that different applications may require various types of SoD. Lastly, most SoD types are informally defined, which creates ambiguity regarding the subjects or specifications [28].

2.5 The proposed SoD deployment in IaaS by $ABAC_{sh}$

Based on the above investigation [22–24, 29], SoD can be defined as an enforcement constraint configured to avoid conflict between policies. This conflict can be due to multi-access requests from different subjects to the same resource simultaneously, or the same subject requesting access to multiple resources at the same time. From this definition, it can be observed that SoD may be viewed as object-operation-oriented, which can be aligned with ABAC's relation between appropriate to enhance SoD by implementing a form of HSoD which will be suitable to be enforced in a dynamic access control policy environment such as ABAC. With RBAC, the centric entity involved in the SoD principle design is the role set. In contrast, ABAC cannot consider a role in the form of an attribute as it can lead to a chaos [30]. Therefore, the focus of this paper regarding formally defining SoD within ABAC will be on attributes and attribute-relations, with no aim to define an application-oriented SoD. Thus, we aim to identify a logical based design for SoD within the ABAC policy model. The proposed work is based on formal logic; exception cases are not encouraged in a formal logic as exceptions make regulations non-monotonic and introduce conflict between proven conclusions [31].

Therefore, the proposed SoD is operation-object orientated that defines a rules-set reflecting the forbidden operations on the set of objects under a specific situation of a collection of entities attributes. Entities include the object, the subject, the environment, and the system context. Moreover, formal logic facilitates SoD rule creation, even by non-expert security administrators. Since the proposed system is attribute-based, it is not necessary to update different locations if a new action restriction is added, deleted, or modified. Object-attributes and operations. We can discern from the above that it is more appropriate to enhance SoD by implementing a form of HSoD which will be suitable to be enforced in a dynamic access control policy environment such as ABAC. With RBAC, the centric entity involved in the SoD principle design is the role set. In contrast, ABAC cannot consider a role in the form of an attribute as it can lead to a chaos [30]. Therefore, the focus of this paper regarding formally defining SoD within ABAC will be on attributes and attribute-relations, with no aim to define an application-oriented SoD. Thus, we aim to identify a logical based design for SoD within the ABAC policy model. The proposed work is based on formal logic; exception cases are not encouraged in a formal logic as exceptions make regulations non-monotonic and introduce conflict between proven conclusions [31]. Therefore, the proposed SoD is operation-object orientated that defines a rules-set rejecting the forbidden operations on the set of objects under a specific situation of a collection of entities attributes. Entities include the object, the subject, the environment, and the system context. Moreover, formal logic facilitates SoD rule creation, even by non-expert security administrators. Since the proposed system is attribute-based, it is not necessary to update different locations if a new action restriction is added, deleted, or modified.

3. An intelligent framework for $ABAC_{sh}$

The framework is designed based on knowledge-agent and it employs rule-based expert system method. This intelligent system is not based on machine learning which will have a percentage of correct answers. This system is based on the available rules; therefore, it is not a type of uncertain approach. The system must guarantee an access decision.

The purpose of this **$ABAC_{sh}$** framework is to prove that AI architecture can contribute in supporting a dynamic access control. In regard to guaranteeing behavior, the followed mechanism in this chapter is based on knowledge available. If there is a shortage in knowledge, the access decision will be denied. There are other AI categories related to uncertain knowledge such as probabilistic reasoning, However, uncertain reasoning is out of this research scope.

3.1 AI scope for the proposed framework

According to [32–34], artificial intelligence systems are designed to think and act. They can be categorized into four types based on the intention of the system: Thinking Humanly, Acting Humanly, Thinking Rationally and Acting Rationally. The category of Thinking Rationally leads to an evolved need for the logic field in artificial intelligence. Involving logic in an intelligent system faces two substantial obstacles. The first one is the difficulty of presenting informal-knowledge using a formal logical notation though the certainty level is less than 100%. The second is that solving problems theoretically is different from solving them practically when the machine capacity is taken into consideration.

The category of Acting Rationally initiates the development of a computer agent. Prior to computer science, the term agent was used in different fields.

Therefore, there are various definitions of agent. However, it can be defined as an entity that acts within an environment by sensing its surroundings to update its knowledge and acts upon that to meet specific goals [35]. The agent function represents an abstract mathematical description, whereas the agent program represents an agent implementation within a physical system.

Problem-solving through an intelligent agent involves four stages. Firstly, the agent formulates its goal. Secondly, it formulates the problem based on five steps: initial state, possible actions, transition model that describes what each action does, goal test and path cost. Thirdly, it searches for a solution by looking for a sequence of actions that leads to the goal. Fourthly, in the execution stage, the solution found is implemented. However, the problem-solving agent is inflexible as each possible state should be hard-coded. Therefore, the complexity of the search stage grows exponentially in relation to the number of states in addition to its inability to infer unobserved information. Therefore, there is a need for logic to reason about the possible states instead of hard-coded all predicted states.

Knowledge-based reasoning is a step in overcoming problem-solving agent limitations. The logic provides a natural language for describing and reasoning about the system. The knowledge-based system is given facts about the external world, and it is asked queries about that world. The rule-based expert system is a popular method that is used to build knowledge-based systems. The rules are used to represent knowledge in the format of IF-THEN. The Inference engine is the reasoning component whereby the system concludes by linking the rules given in the knowledge base with facts supplied from the database. The explanation facilities allow the user to interact with the expert system to get justifications regarding the results produced by the inference engine. Therefore; the AI scope for the proposed intelligent-framework for **ABAC$_{sh}$** is illustrated in **Figure 1**. Modal logic is found to be the most appropriate logic to be used in AI as discussed by [36].

3.2 Logical-based agent architecture

Intelligence security is a fertile approach, as most existing security paradigms suffer from reactive and fragmented approaches [37]. In a frequently changing infrastructure, deploying an agent-based mechanism will be an advantage [38].

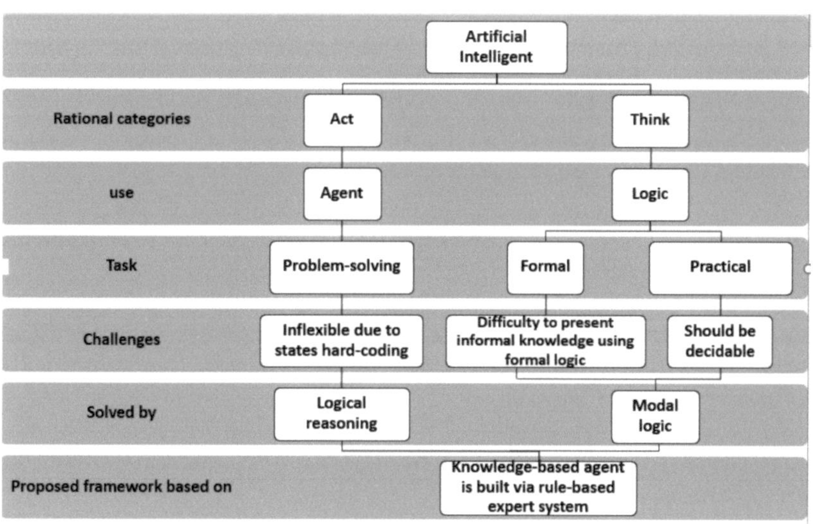

Figure 1.
AI scope for the proposed framework.

Modal logic is a candidate that supports a logical approach in artificial intelligence systems [39]. The main component of a knowledge-agent is a Knowledge-Base (KB) that consists of a set of sentences expressed using formal logic, in addition to two generic functions that involve logical inference. The first function is known as TELL, and adds new sentences (facts) to the KB to provide it with the required information. The second function is known as ASK, and queries the known information from the KB to determine the next step. The process between TELL and ASK will end as soon as the desired action is selected. The interaction between these two generic functions is similar to the updating and querying in databases, as illustrated in **Figure 2**. When an agent program is called upon, it performs two main actions. Firstly, it will TELL the KB what it perceives. Secondly, it ASKs the KB what action should be taken.

Therefore, agent-based architecture is suitable to represent an ABAC model. The logical agent, furthermore, will be appropriate for the proposed modal logic scheme. **Table 1** demonstrates how knowledge-based agent architecture can represent an ABAC system. The logical agent can be designed to represent an access-request state through a process of inference to derive a new representation of the access-request state that can be used to deduce required actions. The proposed access-control logic agent will be founded on knowledge-based agents, as this type of agent is logic-based [34].

3.3 $ABAC_{sh}$ conceptual requirement

Based on the analysis and investigations addressed an analytical study published by this chapter author in [40], the critical requirements in designing an ABAC model are listed below.

- Req.1 ABAC model definition requires to identify the configuration points. Each point should be formalized via the proper languages. The configuration point indicates the necessary configurations to be accomplished via the ABAC model processing for computing the access decision. These points are known as functional points. It is more convenient to minimize the number of configuration points as they affect the system's computational complexity.

- Req.2 ABAC is identity-free. Therefore, identifications such as subject-id are not the main elements in access-decision processing.

- Req.3 Avoid the creation of lists or groups in the design, as ABAC is intended to be flexible and able to cope with large enterprises.

- Req.4 Context-attributes reflect the current system state, whereas environment attributes reflect the fixed system characteristics.

- Req.5 ABAC is a multi-factor decision. Therefore, it enables fine-grained access

- Req.6 There are no predefined privileges for subjects as the privileges are computed after an access request is triggered. Policy rules set in ABAC are specified based on attributes. As a result, the permissible operations will be defined upon access-request.

- Req.7 The two basic functionalities in ABAC are attribute-assignment and rules-creation.

- Req.8 Security principles such as Separation of Duty (SoD) must be enforced.

```
function KB-AGENT(percept) return an action
static: KB, a knowledge base
t, a counter, initially 0, indicating time
TELL(KB,MAKE-PERCEPT-SENTENCE(percept,t))
action←ASK(KB,MAKE-ACTION-QUERY(t))
TELL(KB, MAKE-ACTION-SENTENCE(action,t)
t←t + 1 return action
```

Figure 2.
A generic knowledge-based agent function [34].

Components	Agent architecture	ABAC requirements
knowledge base(KB)	Background sentences	Predefined sets of entities, attributes and policy rules
	To represent action(s)	The action is access-decision
inference system	Infer (i.e arrive to a conclusion via reasoning) hidden properties of the world to add new sentence to KB	New sentences are added each time an access-request is triggered which consist of a combination of attributes with the request operation
	Infer based on the predefined sentences and the new ones to conclude with appropriate actions	Reasoning based on the attributes set and the policy rule-sets to conclude with an appropriate action (allow or deny) the access-request

Table 1.
Mapping knowledge-based agent with ABAC requirements.

The enhanced attribute-based access control $\mathbf{ABAC_{sh}}$ fulfills requirement Req.1 by employing one main configuration point that is ABAC agent. This agent takes as an input, the access request parameters which consist of the subject, the object, and the operation (s, o, opr). Then it returns the access decision that indicates if the subject is allowed to operate on the object or it is denied. Compared to ABAC α , which has four configuration points, the policy configuration here is reduced to one, as the proliferation of policy configuration points can introduce difficulties in policy expression and comprehension [5]. For Req.2, in the Policy Decision Point (PDP), the decision-making process considers the subject attributes in addition to other attributes, instead of depending solely on the subject identity information. In Req.3, grouping is studied by HGABAC [41] to facilitate the addition of a hierarchy feature to ABAC.

However, grouping and listing will impede the flexible nature of ABAC [30, 42]. Therefore, permissions grouping and listing are avoided in this $\mathbf{ABAC_{sh}}$. The decision calculation is based on four sets of attributes: subject-attributes, object-attributes, environment-attributes, and system-context attributes, all of which are taken into consideration in the proposed design to meet requirement Req.4. System context attributes have a special sensor to obtain an up to date system state to meet requirement Req.5. The privilege decision is calculated based on the attributes relation defined in the policy-rules. Therefore; the privilege value is returned after the access-decision is triggered, which meets the requirement Req.6. There are two core functions of $\mathbf{ABAC_{sh}}$. The first function takes place at the initial system stage, where the attribute pairs (name:value) are created for the defined access control system entities

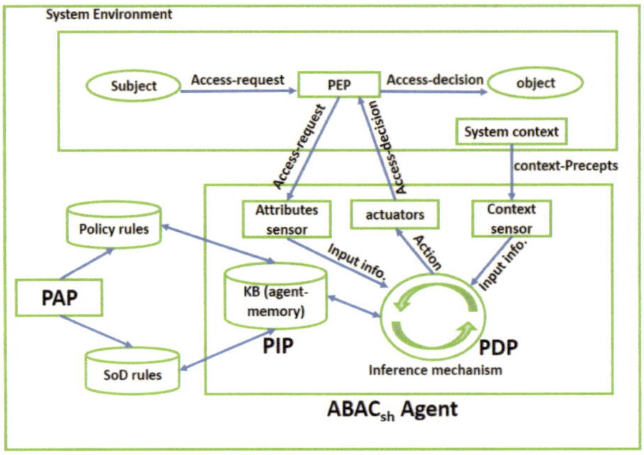

Figure 3.
The proposed intelligent framework for ABAC$_{sh}$.

(subject, object, environment, and context). The second function is rules creation, which represents the SoD-rules and Policy-rules in the form of capability which indicates the access-rights. These two functions meet the ABAC requirement Req.7. An initial SoD is introduced in this design in the type of a DSoD. The elimination of policy-rule conflicts can be achieved by an object-operation oriented constraint. A formal presentation of the proposed SoD enforcement sentences is defined and will be flexible to manage the set of constraints and meet the system requirements Req.8 since the administrator can modify the set of SoD sentences. The proposed SoD will be enforced after the access-request is triggered where an action is forbidden based on a collection of attributes.

3.4 The proposed framework

The proposed intelligent framework for **ABAC$_{sh}$** has been published by this chapter author in [40]. **Figure 3** shows the framework components. The proposed **ABAC$_{sh}$** model framework focuses on three functional points in Ref. to XACML framework: PDP, PIP and PAP. The Policy Enforcement Point (PEP) enforces the access decision. The PDP (Policy Decision Point) involves the core logical reasoning that takes place in an inference mechanism where the access decision is processed. The PAP (Policy Administration Point) involves rule creation by the system administrators. The PIP (Policy Information Point) involves information collection.

4. ABAC$_{sh}$ implementation for IaaS cloud via OpenStack

This section demonstrates the visibility of **ABAC$_{sh}$** in IaaS cloud by introducing an enforcement architecture based on OpenStack. That is followed by a prototype implementation and performance evaluation that illustrates the advantages of the proposed **ABAC$_{sh}$** extension over the existing access control model. This section discuss the following points

- Designing enforcement architecture for **ABAC$_{sh}$** that utilizes telemeter service deployment to be used in feeding Policy Information Point (PIP) with attributes values.

- A prototype implementation of an extended nova access control model with an intelligent ABAC.
    - Extend the nova policy enforcement point (PEP) to communicate with an external policy engine
    - The proposed external policy-engine works as policy decision point (PDP)
- The introduced PDP follows $\mathbf{ABAC_{sh}}$ by
    - Utilizing the attributes in access decision-making process.
    - Involving forward-chaining algorithm that works as logical reasoning for access decision processing.
- Three experiments are studied to compare and contrast the extended $\mathbf{ABAC_{sh}}$ with the default nova-OpenStack access control model.
- The Quality of Service (QoS) measurement is discussed based on response time as a performance metric.

5. OpenStack access control model (OSAC)

A key component in building a virtualization environment is its operation via the hypervisor. The hypervisor on its own cannot build IaaS. Therefore, a cloudstack such as OpenStack, Cloud-Stack or OpenNebul is required. According to the current industry, OpenStack is likely to become a dominant cloud-stack [43]. OpenStack is an open-source cloud computing platform that offers an IaaS layer of service. OpenStack IaaS infrastructure supports agent communication. For example, network nodes in the OpenStack activate a DHCP agent to deploy a DHCP service [44]. OpenStack was selected to be the experimental platform for this research as it has a supportive and active community of both academic researchers and commercial bodies.

OpenStack can deploy different access control models within its infrastructure [45]. For example, nova configuration files can be protected via several implementations such as centralized logging, policy file (policy.json) and MAC framework (Mandatory Access Control). The availability of access control models depends on the hypervisor vendor. The supported models are Mandatory Access Control (MAC), Discretionary Access Control (DAC), and Role-Based Access Control (RBAC).

The Openstack access control model (OSAC) that enables both operators and users to access resources for specific services is a type of RBAC [46]. The keystone [47] supports the notation of roles and groups. Each user should be associated with a group, and each group has a list of roles. For a user to be granted access to a service, Openstack service takes into consideration his/her role, though as the first authorization step, the OpenStack PEP (Policy Enforcement Point) takes into consideration the policy rules associated with the resources before it checks the user role. Therefore, the policy enforcement middleware enables fine-grained access. Each Openstack service defines the access control policies rules for its resources in a specific policy file called policy.json.

5.1 Policy engine

Policy engine in OpenStack is a type of authorization engine that return back a decision based on some policy rules that indicating if a specific operation is allowed or denied [14, 45, 48]. The default policy engine is maintained via Oslo policy, and the access request is issued via API communication. Oslo policy is completely separated from RBAC model [49]. The developer can view Oslo policy rules that are related to nova via the command "oslopolicy-policy-generator {namespace nova". The list of rules verifies if the user credentials are matching to grant access to the requested resources. The user credentials are stored in the format of a token. The token holds information related to the token itself in addition to the user, the project, the domain, the role, the service and the endpoint. The policy rules are stored in JSON (JavaScript Object Notation) file format.

In policy.json, the access policy consists of two main parts "<target>": "<rule>" [47] . The target is known as an action that indicates the API call for an operation such as "start an instance". The rules can be one of the following: allow all, deny all, a special check, a comparison of two values, Boolean expressions based on simpler rules. The special check gives the developers an opportunity to extend the OpenStack policy engine. The special check can indicate, a role that is extracted from token information, or a complex rule by defining an alias, or a target URL that delegates the check to an external policy engine.

5.2 Nova authorisation data-flow

Each service in Openstack has its own access control configuration points which involve PEP, PIP, PDP and PAP. The information ow between nova access-control configuration points is demonstrated in **Figure 4**. In the original Openstack architecture, Nova PEP will send a token that contains the information of the access request to Nova PIP to retrieve the object information. Then Nova PEP sends the information of the subject, object and request to Nova PDP in order for Nova PDP request an access control policy from Nova PAP. Nova PDP evaluate the access request based on the policy and return the access decision to Nova PEP.

5.3 Forward-chaining algorithm

In the search stage of the problem-solving agent, there is a need to use a proper searching algorithm that meets the problem scenario and the input information. The search algorithms that are used in rule-based systems are backward chaining, forward-chaining and a mixture of both of them [34, 51]. **Table 2** compares between the reasoning algorithms which are referred to as chaining in some literature.

Many researchers avoid the Logic Theory Machine, which is based on forward-reasoning due to the computation complexity. However, this complexity is due to the classical mathematical logic and it is not due to the forward-reasoning concept [52]. Classical mathematical logic such as propositional logic and First-order logic. Therefore, the computational complexity of forward-chaining when it is used in nonclassical logic such as deontic logic will be decidable, and it will have an acceptable computation complexity. A simple algorithm for forward-chaining is illustrated in **Figure 5**. Forward reasoning search iteration is based on facts and rules to find a conclusion.

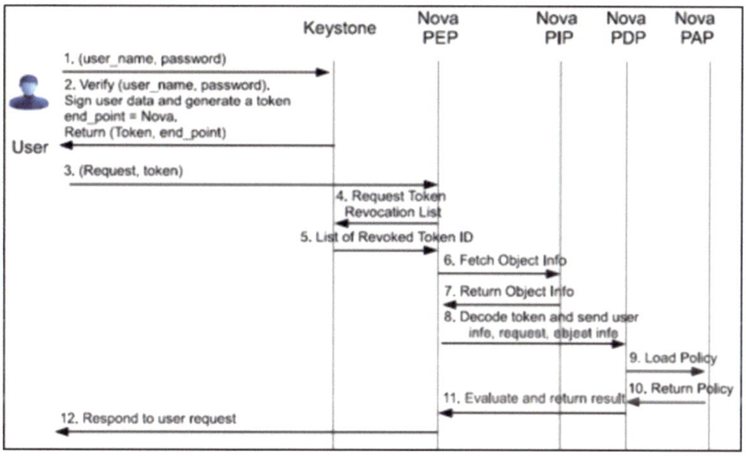

Figure 4.
Nova authorization data-flow [50].

	Forward-chaining	Backward-chaining
Known as	Forward reasoning (Data driven)	Backward reasoning (Goal driven)
Reasoning start with	A Set of facts to reach a goal (or hypothesis)	A hypothesis (goal) to reach the facts behind it
When applicable	If the goal is unknown	If the set of goals are known

Table 2.
Comparing forward-chaining with backward-chaining.

```
newFacts=False
For rule in rule-list
     if all premises match fact-base:
          For each fact in consequences:
               if fact not in fact-base:
                    add fact to fact-base
                    new Facts = True
If newFacts: goto 1
```

Figure 5.
Forward chaining algorithm.

5.4 ABAC implementation in OpenStack

5.4.1 Enforcement architecture

The core characteristics of $\mathbf{ABAC_{sh}}$ are to work as an intelligent agent that sense the attributes (the environment, the system context, the subject and the object) in order to search for an access decision using forward chaining (forward-reasoning). The set of attributes represent facts whereas the set of policy rules represent the rules.

The proposed **ABAC$_{sh}$** enforcement architecture employs the Telemetry service of OpenStack. The telemetry service in Openstack provides the facility to sense the IaaS cloud for environment attributes and the context attributes. The Telemetry service facilitates polling information from the computing service since the proposed access control agent **ABAC$_{sh}$** will use the collected information for the attributes assignment process. As an example, the nova service access control process will be used to illustrate **ABAC$_{sh}$** extension. Section 4.3 introduces the default data ow of nova service access control while **Figure 6** illustrates nova service access control with **ABAC$_{sh}$** extension. The proposed **ABAC$_{sh}$** enforcement architecture focuses on three configuration points: PIP (Policy Information Point), PAP (Policy Admission Point) and PDP (Policy Decision Point). **ABAC$_{sh}$** enforcement architecture is divided into three components as follows:

1. **ABAC$_{sh}$** PIP: this is used to collect attributes information from the access control entities, the environment, and the system context. PIP can be achieved in Openstack through configuring the Telemetry component. The Telemetry service is designed to support a billing system by gathering the required information. Therefore, its structure will be beneficial in providing PIP with required attribute information. Telemetry consists of five building blocks: Compute Agent, Central Agent, Collector, Data Store and API Server in order to perform five essential functions [53]. **Figure 7** summarizes the Telemeter process to collect data for further analysis. Telemeter can be configured to collect the attributes and save them in JSON file as this file format is used to store policy rules in OpenStack

2. **ABAC$_{sh}$** PAP: The knowledge database for **ABAC$_{sh}$** model consists of access rules from SoD rules and Policy rules. The access rules are created by the system administrator. Those rules will be stored in JSON file format to facilitate its implementation in OpenStack.

3. **ABAC$_{sh}$** PDP: this is the logical component which reasons about access control in **ABAC$_{sh}$**. **ABAC$_{sh}$** PDP will get an access-request sentence from **ABAC$_{sh}$** PEP that consist of the attributes information with the access request. **ABAC$_{sh}$** PDP will load the access rules from **ABAC$_{sh}$** PAP. **ABAC$_{sh}$** PDP accomplishes logical reasoning through forward-chaining algorithm. The result of the logical reasoning indicates if the access is permitted or denied.

5.4.2 Prototype implementation

The first stage of **ABAC$_{sh}$** deployment in Openstack is to be implemented on nova component. **ABAC$_{sh}$** PDP part will be implemented as a prototype.

- **Scope and Assumption.**

  IaaS access control tenant scope can be a single tenant [5], multi-tenant [54–56] and collaborating parties a cross-clouds [57, 58]. The implementation scope of access control in this chapter is within single tenant whereas its hypothesis is applicable to multi-tenant and cross-clouds as the big concept behind **ABAC$_{sh}$** is user-id free and attributes-based. The proposed **ABAC$_{sh}$** is not replacing OpenStack RBAC in this stage. Instead, it allows fine-grained access control and opens prospective avenues to replace RBAC in the near feature.

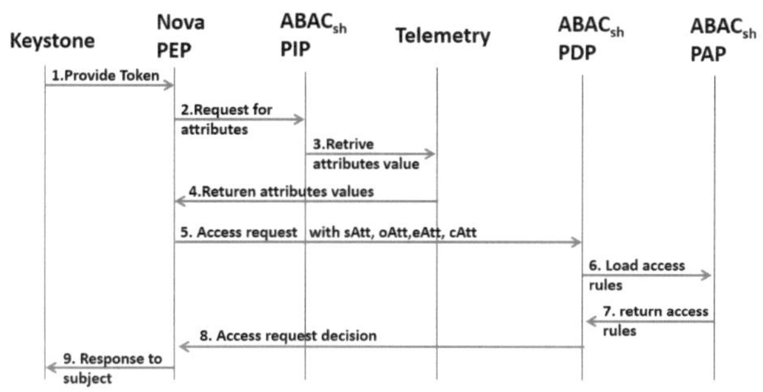

Figure 6.
Proposed $ABAC_{sh}$ for Openstack nova.

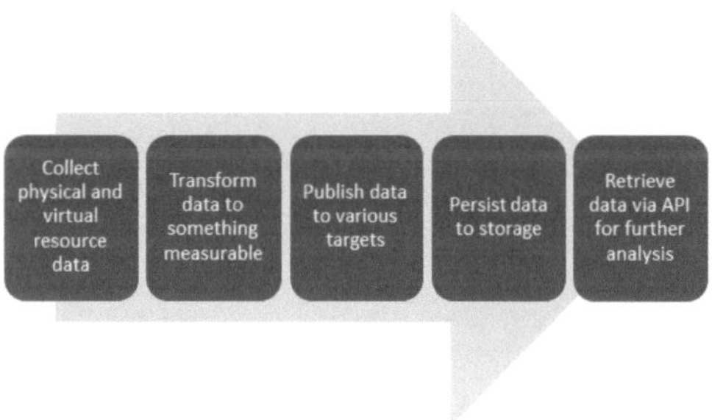

Figure 7.
Telemeter process.

- **OpenStack Testbed**.

OpenStack aids in deploying IaaS cloud. **Figure 8** shows the deployed testbed in this chapter. It is installed in three machines using Ubuntu 16.04 LTS as an operating system and OpenStack Ocata the latest release (Feb2017). One machine is configured as a controller which provides OpenStack main server in addition to networking services (neutron), keystone, nova and glance. The Two other machines are configured as compute nodes where virtual machines are hosted. The machines specification is Intel Core i5–4460 CPU Processor 3.20GHz _ 4, 15.5 GiB memory, 235 GB Disk and 2 NICs cards. The testbed networking consists of two LANs: management network and data network. The management network traffics the Openstack service communication where data network connects the communication of the virtual machine. This IaaS is a private cloud where OpenStack services and the VMs are accessed by the LAN users.

- **Data flow**.

Nova policy engine is embedded within its configuration files, therefore it is considered as one of OpenStack's limitations. However, the default policies can be overwritten if policy.json is enabled. Policy.json can be configured to call an external policy engine through URL. The token hold information that can be passed from OpenStack keystone to $\mathbf{ABAC_{sh}}$ policy engine via RESET GET-call. Nova PEP receives an access decision from $\mathbf{ABAC_{sh}}$ policy engine via RESET POST-call. $\mathbf{ABAC_{sh}}$ policy engine use a forward-chaining algorithm to produce an access control decision. The access control reasoning takes facts which are subject and object attributes, in addition to the system and context attributes.

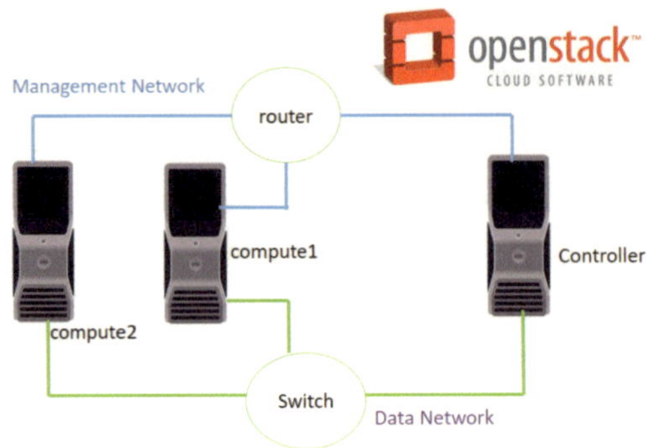

Figure 8.
OpenStack testbed in physical machines.

Based on access rules defined by the administrator, the access request will be allowed or denied. **Figure 9** shows the **ABAC$_{sh}$** PDP extension to nova authorization. A policy engine is designed and implemented to add an access control enforcement based on attributes.

In access control terminology, the Openstack users are the subject, the nova resources are the objects, the policy.json is PEP and **ABAC$_{sh}$** policy engine is the PDP. The attributes are extracted from the following channels

- The subject attributes can be extracted from keystone token where the available information is user name, user id, user passward, role id, role name. The Token information can be retrieved from "content-type:application/json' 'through curl command.

- The extracted nova metrics from the OpenStack system via a command openstack quota show are considered as the object attributes. The attributes information is stored in JSON _le format

- The nova environment attributes are extracted via the OpenStack command openstack hypervisor stats show. The attributes information is stored in JSON _le format

- The context attributes are not implemented in this prototype but it is visible to be included via Telemetry OpenStack service

  ABAC$_{sh}$ policy engine server is implemented using several programming technologies. The web server is developed using Python programming language with web.py since OpenStack services is using python. RESETful API utilities are used to allow the communication between **ABAC$_{sh}$** and OpenStack APIs. The forward reasoning function is programmed using java since this programming language can be smoothly integrated into web programming. To allow the technical interaction between python and java, jpype is used [59, 60]. Data is stored in JSON file formats such as policy data and attributes data.

5.4.3 Performance evaluation

The aim of this performance evaluation is to detect if **ABAC$_{sh}$** deployment in Openstack introduces any significant overhead. The efficiency of deploying an access control model depends on several factors. The quality of service (QoS) measures

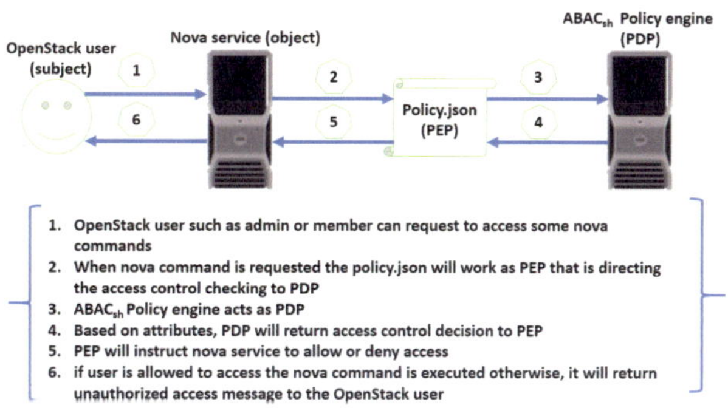

Figure 9.
The prototype implementation data flow.

can be calculated by performance properties and computation complexity [61, 62]. In this section, the performance metrics are evaluated. The performance metrics consist of four elements: response time, policy repository and retrieval, policy distribution, and Integrated with authentication function [62]. **Table 3** explains theses performance metrics elements and on which access control components they can be applied. Since the implemented prototype is the **ABAC$_{sh}$** PDP, the followed experiments will measure response time. With regard to policy repository and retrieval, the implemented **ABAC$_{sh}$** use JSON file to store access control policy which does not add any extra hardware or software cost to OpenStack IaaS cloud as this form of policy storage is used by OpenStack. The remaining two performance metrics elements are not calculated in this stage as PIP is not implemented.

5.4.4 Experiment content

In this section, the performance evaluation of the implemented **ABAC$_{sh}$** prototype in OpenStack is presented. Specifically, **ABAC$_{sh}$** policy engine which represents PDP of access control model is discussed. The experiments fall into three parts where the response time is calculated. Response time indicates the time consumed by the system in order to process the access request decision call. The response time has been used to measure the performance in several OpenStack implementations such as in [63, 64]. In these experiments, OpenStack cloud was installed in physical servers running Ubuntu 16.04 LTS release. Three types of execution time can be measured [65, 66]. The first one is real time that reflects the wall clock where the time is calculated from the start till the end of the call including the waiting time and time used by other processes. The second one is user-time that reflects the actual CPU-time spent outside the kernel during the process call in user-mode without considering other processes. The third one is sys-time that reflects the actual CPU-time spent within the kernel during the process execution.

Three experimental settings have been implemented as explained below.

- Experiment 1 (Exp1): The response time for the default access control model to process access request to nova resources. The default use RBAC and Oslo policy engine.

- Experiment 2 (Exp2): The response time for extending the default nova policy engine with **ABAC$_{sh}$** that utilizes 24 attributes in access control processing

- Experiment 3 (Exp3): The response time for extending the default nova policy engine with **ABAC$_{sh}$** that use forward-chaining for access control reasoning.

5.4.5 Experimental results and discussion

Each experiment was run five times, and then the average value was recorded. Five scenarios were observed by increasing the number of requests from five to twenty-five as illustrated in **Table 4**. The request indicates an access control request from a user (subject) to access nova-resources (object).

Based on usability engineering [67, 68] The response time value can be within three categories: over 0.1 seconds will give the user a feeling that the system is reacting instantaneously, over 1.0 seconds will give the user a feeling of a delay but

Performance metrics element	Description	The applicable Access Control Component
Response time	The time required to process access request should meet the organization requirement	PEP, PDP, PIP, PRP
Policy distribution	If there exist a mechanism that can be used for access control policy distribution	PAP, PIP, PRP
Integrated with authentication function	If the subject and object can be associated with some identifications through an authentication function.	PIP
Key: Policy Decision Point (PDP), Policy Administration Point (PAP), Policy Enforcement Point (PEP), Policy Information Point (PIP), Policy Retrieval Point (PRP)		

Table 3.
Performance evaluation metrics.

No. of Requests	Response time	Exp1	Exp2	Exp3
5	Real	8.67	8.96	8.77
	User	5.56	5.52	5.55
	System	0.42	0.44	0.40
10	Real	17.24	17.34	17.24
	User	10.99	11.05	11.11
	System	0.88	0.86	0.83
15	Real	25.52	26.08	25.87
	User	16.46	16.61	16.64
	System	1.25	1.31	1.22
20	Real	34.40	34.56	34.45
	User	22.07	22.17	22.23
	System	1.68	1.66	1.56
25	Real	43.02	43.05	43.07
	User	27.64	27.63	27.82
	System	2.09	2.10	1.96

Table 4.
Experimental results.

will stay uninterrupted, over 10 seconds the user will lose his/her attention and will search for something to work on till the computer responds.

Three time values has been recorded as illustrated in **Table 4**: real-time, user-time and sys-time. In this study, real-time and sys-time have a direct reflect on the performance analysis whereas user-time is reflecting the processing outside the kernel. The real-time shows the access control execution time in additions to the other OpenStack cloud processes that introduce some delay by blocking the process or introducing a waiting time. Therefore, this measurement will indicate the effect of our extended $\mathbf{ABAC_{sh}}$ nova to the overall OpenStack system.

The graph in **Figure 10** compares the real-time for the three set of experiments. The increase is 0.05 seconds when the extended $\mathbf{ABAC_{sh}}$ nova employ forward reasoning in access decision processing as shown in **Table 5** while the increase is 0.145 seconds when $\mathbf{ABAC_{sh}}$ uses twenty-four attributes in access decision processing. Therefore, there is an increase of 0.56% when attributes are added to the policy engine and 0.19% when the forward-chaining algorithm is added. Consequently, the increase in response time is negligible in Ref. to the usability engineering when the nova default access control is extended with part of the proposed ABAC enhancement.

On the other hand, sys-time gives the process execution only within the kernel regardless of the other tasks. Therefore, the time for the 25 requests dropped from 43.02 seconds within real-time to 2.09 seconds within sys-time during Exp1 which involve default nova access control. The sys-time comparison for the three experiments is illustrated in **Figure 11**. The results show a slightly better performance of

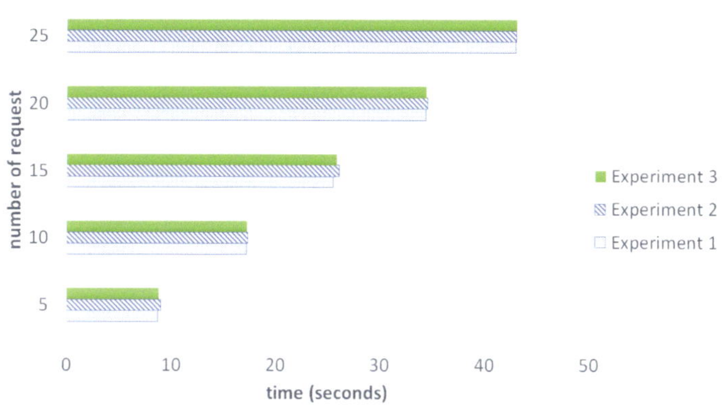

Figure 10.
Real-time for access control processing in nova.

Experiment 2 - Experiment 1	Experiment 3 - Experiment 1
0.29	0.1
0.1	0
0.56	0.35
0.16	0.05
0.03	0.05
Average	Average
0.145	0.05
Percentage	Percentage
0.56%	0.19%

Table 5.
Comparing real-time values.

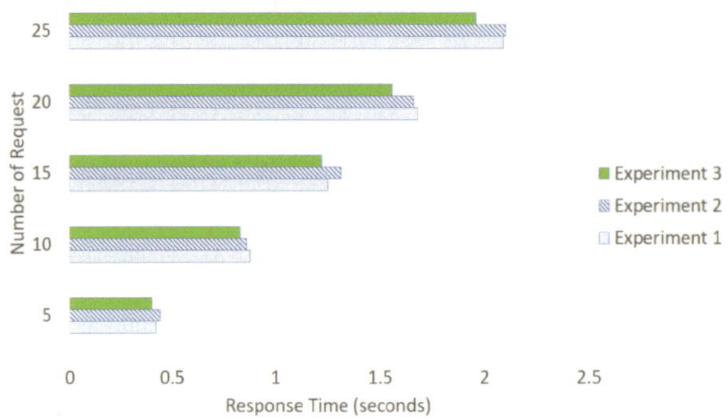

Figure 11.
Sys-time for access control processing in nova.

Experiment 2 - Experiment 1	Experiment 3 - Experiment 1
0.02	−0.02
−0.02	−0.05
0.06	−0.03
−0.02	−0.12
0.01	−0.13
Average	Average
0.05	−0.07
Percentage	Percentage
4%	−5.5%

Table 6.
Comparing sys-time values.

5.5% for extending the default nova access control when forward-reasoning has been utilized whereas an increase of 4% over the default nova when 24-attributes are used in the policy-engine as illustrated in **Table 6**.

From these results, the $\mathbf{ABAC_{sh}}$ shows an acceptable performance compared to the default OpenStack access control within nova service. This section demonstrates the enhanced attribute-based access control $\mathbf{ABAC_{sh}}$ performance improvement when attributes and forward reasoning algorithm are employed. It has been noticed that the performance improvement is liner in **Figure 10** when only attributes are involved in access decision. Whereas in **Figure 11** when forward reasoning is involved, an improvement in performance has been noticed. This indicates an opportunity of enhancing the IaaS-cloud security when logical reasoning and AI mechanism are involved in access control system.

5.4.6 Experiments limitations

The main aim of the experiments in this chapter is to study the performance improvement when attribute-based access control model is introduced into IaaS cloud. The experiment scale is limited to a private cloud in a LAN set-up. Therefore, the network performance metrics has not been studied such as the latency and throughput. The implementation in this chapter does not involve the PIP component of the access control, therefore only a simple forward reasoning algorithm has been deployed without knowledge update component. The used database for knowledge is written manually whereas the system should use an automated information collection method if PIP is implemented. One subject is involved in each experiment, therefore multi-access has not been investigated in this chapter. Multi-tenant study is a critical future work.

6. Conclusions

This Chapter is focusing on the problem of deploying access control in a dynamic environment. Access control is one of the information security principles where the system user access is controlled through an access policy. In the cyber-security world where systems and devices are distributed in different locations, there is a need to have an access control model that is able to cope with a dynamic environment where new users with different privileges are joining and leaving the system.

This chapter is proposing to deploy an enhanced version of attribute-based access-control named $\mathbf{ABAC_{sh}}$. This model is deploying knowledge base category of AI. A proof of concept is implemented in the cloud computing environment to measure the performance and the visibility of such a deployment.

Author details

Shadha Mohamed Sulaiyam ALAmri
University of Technology and Applied Sciences, Muscat, Oman

*Address all correspondence to: shadha-alamri@hct.edu.om

References

[1] W. E. Forum, "Fourth Industrial Revolution Beacons of Technology and Innovation in Manufacturing," Geneva, 2019. [Online]. Available: http://www3.weforum.org/docs/WEF_4IR_Beacons_of_Technology_and_Innovation_in_Manufacturing_report_2019.pdf.

[2] J. Qiu, Z. Tian, C. Du, Q. Zuo, ... S. S.-I. I. of T., and U. 2020, "A survey on access control in the age of internet of things," *ieeexplore.ieee.org*, vol. 7, no. 6, pp. 4682-4696, 2020, doi: 10.1109/JIOT.2020.2969326.

[3] Y.-G. Kim, C.-J. Mon, D. Jeong, J.-O. Lee, C.-Y. Song, and D.-K. Baik, "Context-Aware Access Control Mechanism for Ubiquitous Applications," Springer Berlin Heidelberg, 2005, pp. 236-242.

[4] Y.-G. Kim and Y. Lee, "Context Information-based Application Access Control Model," in *Proceedings of the 10th International Conference on Ubiquitous Information Management and Communication - IMCOM '16*, 2016, pp. 1-5, doi: 10.1145/2857546.2857623.

[5] X. Jin, "Attribute-based access control models and implementation in cloud infrastructure as a service," THE UNIVERSITY OF TEXAS AT SAN ANTONIO, 2014.

[6] A. Cavoukian, M. Chibba, G. Williamson, and A. Ferguson, "The Importance of ABAC: Attribute-Based Access Control to Big Data: Privacy and Context," *Priv. Big Data Institute, Ryerson Univ. Toronto, Canada*, 2015, Accessed: May 17, 2016. [Online]. Available: http://www.ryerson.ca/content/dam/pbdi/Resources/The Importance of ABAC to Big Data 05-2015.pdf.

[7] F. Li, "Context-Aware Attribute-Based Techniques for Data Security and Access Control in Mobile Cloud Environment." Apr. 01, 2015.

[8] K. Nahrstedt and R. Campbell, "Security for Cloud Computing," 2012. Accessed: Oct. 08, 2015. [Online]. Available: https://illinois.edu/blog/files/695/66281/2737.pdf.

[9] X. Yao, X. Han, and X. Du, "A lightweight access control mechanism for mobile cloud computing," in *2014 IEEE Conference on Computer Communications Workshops (INFOCOM WKSHPS)*, Apr. 2014, pp. 380-385, doi: 10.1109/INFCOMW.2014.6849262.

[10] D. Al Kukhun, "Steps towards adaptive situation and context-aware access: a contribution to the extension of access control mechanisms within pervasive information systems," Research Institute in Computer Science of Toulouse, 2012.

[11] Z. Liu and J. Wang, "A fine-grained context-aware access control model for health care and life science linked data," Multimed. Tools Appl., vol. 75, no. 22, pp. 14263-14280, Jan. 2016, doi: 10.1007/s11042-016-3269-6.

[12] K. K. Venkatasubramanian, T. Mukherjee, and S. K. S. Gupta, "CAAC -- An Adaptive and Proactive Access Control Approach for Emergencies in Smart Infrastructures," ACM Trans. Auton. Adapt. Syst., vol. 8, no. 4, pp. 1-18, Jan. 2014, doi: 10.1145/2555614.

[13] C. Choi, J. Choi, and P. Kim, "Ontology-based access control model for security policy reasoning in cloud computing," J. Supercomput., vol. 67, no. 3, pp. 711-722, 2014, doi: 10.1007/S11227-013-0980-1.

[14] S. Bhatt, F. Patwa, and R. Sandhu, "An Attribute-Based Access Control Extension for OpenStack and Its Enforcement Utilizing the Policy Machine," in *2016 IEEE 2nd International Conference on Collaboration and Internet Computing (CIC)*,

Nov. 2016, pp. 37-45, doi: 10.1109/CIC.2016.019.

[15] R. L. Krutz and R. D. Vines, *Cloud security: a comprehensive guide to secure cloud computing*. Wiley Pub, 2010.

[16] W. Li, H. Wan, X. Ren, and S. Li, "A Refined RBAC Model for Cloud Computing," in *2012 IEEE/ACIS 11th International Conference on Computer and Information Science*, May 2012, pp. 43-48, doi: 10.1109/ICIS.2012.13.

[17] D. F. D. Ferraiolo, R. Sandhu, S. Gavrila, D. R. Kuhn, and R. Chandramouli, "Proposed NIST standard for role-based access control," ACM Trans. Inf. Syst. Secur., vol. 4, no. 3, pp. 224-274, Aug. 2001, doi: 10.1145/501978.501980.

[18] D. Nguyen, J. Park, and R. Sandhu, "A provenance-based access control model," in *2012 Tenth Annual International Conference on Privacy, Security and Trust*, Jul. 2012, pp. 137-144, doi: 10.1109/PST.2012.6297930.

[19] D. Ferraiolo, V. Atluri, and S. Gavrila, "The Policy Machine: A novel architecture and framework for access control policy specification and enforcement," J. Syst. Archit., vol. 57, no. 4, pp. 412-424, 2011, doi: 10.1016/j.sysarc.2010.04.005.

[20] R. T. Simon and M. E. Zurko, "Separation of duty in role-based environments," in *Proceedings 10th Computer Security Foundations Workshop*, 1997, pp. 183-194, doi: 10.1109/CSFW.1997.596811.

[21] C. T. Hu, D. F. Ferraiolo, and D. R. Kuhn, *Assessment of Access Control Systems*. US Department of Commerce, National Institute of Standards and Technology, 2006.

[22] D. Servos and S. L. Osborn, "Current Research and Open Problems in Attribute-Based Access Control," ACM Comput. Surv., vol. 49, no. 4, pp. 1-45, Jan. 2017, doi: 10.1145/3007204.

[23] K. Z. Bijon, "Constraints for attribute based access control with application in cloud IaaS," THE UNIVERSITY OF TEXAS AT SAN ANTONIO, 2015.

[24] D. Nguyen, "Provenance-based access control models," The University of Texas at San Antonio, 2014.

[25] D. Ferraiolo, R. Chandramouli, R. Kuhn, and V. Hu, "Extensible Access Control Markup Language (XACML) and Next Generation Access Control (NGAC)," in *Proceedings of the 2016 ACM International Workshop on Attribute Based Access Control - ABAC '16*, 2016, pp. 13-24, doi: 10.1145/2875491.2875496.

[26] R. A. Botha and J. H. P. Eloff, "Separation of duties for access control enforcement in workflow environments," IBM Syst. J., vol. 40, no. 3, pp. 666-682, 2001, doi: 10.1147/sj.403.0666.

[27] P. Biswas, R. Sandhu, and R. Krishnan, "Label-Based Access Control," in *Proceedings of the 2016 ACM International Workshop on Attribute Based Access Control - ABAC '16*, Mar. 2016, pp. 1-12, doi: 10.1145/2875491.2875498.

[28] V. D. Gligor, S. I. Gavrila, and D. Ferraiolo, "On the formal definition of separation-of-duty policies and their composition," in *Proceedings. 1998 IEEE Symposium on Security and Privacy (Cat. No.98CB36186)*, 1998, pp. 172-183, doi: 10.1109/SECPRI.1998.674833.

[29] S. Jha, S. Sural, V. Atluri, and J. Vaidya, "Enforcing Separation of Duty in Attribute Based Access Control Systems," in *Information Systems Security*, Springer, Cham, 2015, pp. 61-78.

[30] E. Coyne and T. R. Weil, "ABAC and RBAC: Scalable, Flexible, and Auditable Access Management," IT Prof., vol. 15, no. 3, pp. 14-16, May 2013, doi: 10.1109/MITP.2013.37.

[31] N. Dinesh, A. Joshi, I. Lee, and O. Sokolsky, "Permission to speak: A logic for access control and conformance," J. Log. Algebr. Program., vol. 80, no. 1, pp. 50-74, 2011, doi: 10.1016/j.jlap.2009.12.002.

[32] R. Fagin, J. Y. Halpern, Y. Moses, and M. Y. Vardi, *Reasoning about knowledge*. MIT Press, 2003.

[33] M. Negnevitsky, *Artificial intelligence: a guide to intelligent systems*, 3rd ed. Addison Wesley/Pearson, 2011.

[34] S. J. (Stuart J. Russell, P. Norvig, and E. Davis, *Artificial intelligence: a modern approach*. Prentice Hall, 2010.

[35] A. J. Soroka, "Agent-based System for Knowledge Acquisition and Management Within a Networked Enterprise," in *Artificial Intelligence Techniques for Networked Manufacturing Enterprises Management*, Springer London, 2010, pp. 43-86.

[36] R. Mastop, "Modal Logic for Artificial Intelligence," 2012, Accessed: Mar. 11, 2017. [Online]. Available: http://www.phil.uu.nl/~rumberg/infolai/Modal_Logic.pdf.

[37] R. Knights and E. Morris, "Move to intelligence-driven security," Netw. Secur., vol. 2015, no. 8, pp. 15-18, Aug. 2015, doi: 10.1016/S1353-4858(15)30071-4.

[38] F. Doelitzscher, C. Reich, M. Knahl, A. Passfall, and N. Clarke, "An agent based business aware incident detection system for cloud environments," J. Cloud Comput. Adv. Syst. Appl., vol. 1, no. 1, p. 9, Jul. 2012, doi: 10.1186/2192-113X-1-9.

[39] M. Huth and M. Ryan, *Logic in computer science: modelling and reasoning about systems*. Cambridge University Press, 2004.

[40] S. M. Sulaiyam Al Amri, "IaaS-cloud security enhancement: An intelligent attribute-based access control framework," in *2018 Majan International Conference (MIC)*, Mar. 2018, pp. 1-9, doi: 10.1109/MINTC.2018.8363159.

[41] D. Servos and S. L. Osborn, "HGABAC: Towards a Formal Model of Hierarchical Attribute-Based Access Control," in *Foundations and Practice of Security*, Springer, Cham, 2015, pp. 187-204.

[42] V. C. Hu *et al.*, "Guide to Attribute Based Access Control (ABAC) Definition and Considerations," 2013. Accessed: May 14, 2016. [Online]. Available: http://www.itbusinessedge.com/itdownloads/security/guide-to-attribute-based-access-control-abac-definition-and-considerations.html.

[43] W. Huang, A. Ganjali, B. H. Kim, S. Oh, and D. Lie, "The State of Public Infrastructure-as-a-Service Cloud Security," ACM Comput. Surv., vol. 47, no. 4, pp. 1-31, Jun. 2015, doi: 10.1145/2767181.

[44] OpenStack, "OpenStack Docs: Configure neutron agents," 2017. https://docs.openstack.org/admin-guide/networking-config-agents.html (accessed Apr. 02, 2017).

[45] OpenStack, "OpenStack Docs: OpenStack Security Guide," 2017. https://docs.openstack.org/security-guide/ (accessed Jul. 28, 2017).

[46] X. Wen, G. Gu, Q. Li, Y. Gao, and X. Zhang, "Comparison of open-source cloud management platforms: OpenStack and OpenNebula," in *2012 9th International Conference on Fuzzy Systems and Knowledge Discovery*

(FSKD), May 2012, pp. 2457-2461, doi: 10.1109/FSKD.2012.6234218.

[47] OpenStack.org, "OpenStack Docs: The policy.json file," 2017. https://docs.openstack.org/ocata/config-reference/policy-json-file.html (accessed Aug. 06, 2017).

[48] A. Young, "Dynamic Policy for Access Control," OpenStack Summit May 2015 Vancouver, 2015. https://openstacksummitmay2015vancouver.sched.com/event/2qcK/dynamic-policy-for-access-control (accessed Jul. 29, 2017).

[49] OpenStack.org, "OpenStack Docs: oslo.policy," 2017. https://docs.openstack.org/oslo.policy/latest/ (accessed Aug. 06, 2017).

[50] X. Jin, R. Krishnan, and R. Sandhu, "Role and Attribute Based Collaborative Administration of Intra-Tenant Cloud IaaS," in *Proceedings of the 10th IEEE International Conference on Collaborative Computing: Networking, Applications and Worksharing*, 2014, pp. 261-274, doi: 10.4108/icst.collaboratecom.2014.257591.

[51] R. Haemmerlé and Rémy, "On Combining Backward and Forward Chaining in Constraint Logic Programming," in *Proceedings of the 16th International Symposium on Principles and Practice of Declarative Programming - PPDP '14*, 2014, pp. 213-224, doi: 10.1145/2643135.2643144.

[52] J. Cheng, S. Nara, and Y. Goto, "FreeEnCal: A Forward Reasoning Engine with General-Purpose," in *Knowledge-Based Intelligent Information and Engineering Systems*, Berlin, Heidelberg: Springer Berlin Heidelberg, 2007, pp. 444-452.

[53] OpenStack.org, "OpenStack Docs: Telemetry service overview," *OpenStack.org*, 2017. https://docs.openstack.org/mitaka/install-guide-rdo/common/get_started_telemetry.html (accessed Aug. 16, 2017).

[54] C. Ngo, Y. Demchenko, and C. de Laat, "Multi-tenant attribute-based access control for cloud infrastructure services," *J. Inf. Secur. Appl.*, vol. 27, pp. 65-84, Dec. 2015, doi: 10.1016/j.jisa.2015.11.005.

[55] D. J. Buehrer and C.-Y. Wang, "CA-ABAC: Class Algebra Attribute-Based Access Control," in *2012 IEEE/WIC/ACM International Conferences on Web Intelligence and Intelligent Agent Technology*, Dec. 2012, pp. 220-225, doi: 10.1109/WI-IAT.2012.268.

[56] N. Pustchi and R. Sandhu, "MT-ABAC: A Multi-Tenant Attribute-Based Access Control Model with Tenant Trust," in *International Conference on Network and System Security*, 2015, pp. 206--220, Accessed: May 24, 2016. [Online]. Available: http://link.springer.com/chapter/10.1007/978-3-319-25645-0_14.

[57] D. Lin, P. Rao, E. Bertino, N. Li, and J. Lobo, "Policy decomposition for collaborative access control," in *Proceedings of the 13th ACM symposium on Access control models and technologies - SACMAT '08*, Jun. 2008, p. 103, doi: 10.1145/1377836.1377853.

[58] P. Rao, D. Lin, E. Bertino, N. Li, and J. Lobo, "An algebra for fine-grained integration of XACML policies," in *Proceedings of the 14th ACM symposium on Access control models and technologies - SACMAT '09*, Jun. 2009, p. 63, doi: 10.1145/1542207.1542218.

[59] S. Menard and L. Nell, "JPype documentation — JPype 0.6.2 documentation," 2014. https://jpype.readthedocs.io/en/latest/ (accessed Aug. 05, 2017).

[60] Nullege, "jpype - Nullege Python Samples." http://nullege.com/codes/search/jpype (accessed Aug. 05, 2017).

[61] Y. A. Younis, K. Kifayat, and M. Merabti, "A novel evaluation criteria to cloud based access control models," in *2015 11th International Conference on Innovations in Information Technology (IIT)*, Nov. 2015, pp. 68-73, doi: 10.1109/INNOVATIONS.2015.7381517.

[62] V. Hu and K. Kent, *Guidelines for access control system evaluation metrics*. 2012.

[63] A. Corradi, M. Fanelli, and L. Foschini, "VM consolidation: A real case based on OpenStack Cloud," Futur. Gener. Comput. Syst., vol. 32, pp. 118-127, Mar. 2014, doi: 10.1016/j.future.2012.05.012.

[64] B. Tang and R. Sandhu, "Extending openstack access control with domain trust," *Netw. Syst. Secur.*, 2014, Accessed: Mar. 17, 2016. [Online]. Available: http://link.springer.com/chapter/10.1007/978-3-319-11698-3_5.

[65] D. MacKenzie, "Ubuntu Manpage: time - run programs and summarize system resource usage," *ubuntu.com*, 2010. http://manpages.ubuntu.com/manpages/xenial/man1/time.1.html (accessed Aug. 10, 2017).

[66] askubuntu.com, "command line - How can I measure the execution time of a terminal process? - Ask Ubuntu," *askubuntu.com*, 2011. https://askubuntu.com/questions/53444/how-can-i-measure-the-execution-time-of-a-terminal-process (accessed Aug. 10, 2017).

[67] J. Nielsen, *Usability engineering*. Academic Press, 1993.

[68] B. Taylor, A. K. Dey, D. Siewiorek, and A. Smailagic, "Using Crowd Sourcing to Measure the Effects of System Response Delays on User Engagement," in *Proceedings of the 2016 CHI Conference on Human Factors in Computing Systems - CHI '16*, 2016, pp. 4413-4422, doi: 10.1145/2858036.2858572.

Chapter 4

Adapting Disruptive Applications in Managing Quality Control Systems in Intelligence Manufacturing

Amusan Lekan, Clinton Aigbavboa, Moses Emetere and James Owolabi

Abstract

Controlling quality has become a major trend in the circle of manufacturers and production managers that engage in intelligent manufacturing all over the world, on account of industry 4.0, in recent times. Intelligent manufacturing therefore is the use of advanced applications, analytics, sensors and Internet of Things (IoT) to improve manufacturing. The aim of the study is to carry out a study on application of disruptive application in managing quality system in intelligent manufacturing with a view to improving manufacturing process in organizations. Survey methods was used in collating responses from production managers of manufacturing companies at selected locations censoring production managers and supervisors on some parameters such as areas of disruptions in the quality assurance monitoring and calibration in production process, issues and challenges involved in quality control systems in manufacturing, Man-Whitney U Test, T-test, Pearson's Test were used to analyze the collated data. Also, this study presents advanced analytical tools and applications to improve quality in manufacturing process. The study finally presents areas of disruptions in the quality assurance monitoring and calibration in production process, issues and challenges involved in quality control systems in manufacturing, emerging areas of application and recommendation for improvement.

Keywords: quality, system, intelligence, adaptation, disruption, manufacturing, process

1. Introduction

Quality control issue is one of the cardinal factors in product manufacturing. It entails profiling areas where quality is to be maintained and enforced. In manufacturing parlance, quality could be described as process and protocol of the needful to ensure that the product is maintained at highest peak value. However, quality has been a watch factor that has resulted in industrial productivity birthed by technological innovation. [1] submitted that defining quality from different perspectives depends on the individual philosophical point of view. Whichever perspective adopted among several perspectives available, the authentic definition is the one that was premised around definition of American Society of Quality. American Society of Quality defines quality as an embodiment of totality of manufacturing essence which is positioned to satisfy an implied need or consumers' product need and expectation [1]. Also, [2] viewed quality as an identity of production essence while [3] described quality as collection of different hall-mark of production optimizations, expressed in implementable units. Therefore, it worth a while controlling the process that lead to formulation of quality, then controlling cost at different stage of production would be very easy. The important nature of quality control in industrial manufacturing therefore necessitates the establishment of control system during industrial production process.

Quality control system according to [3, 4] ensures emergence of quality product as output of manufacturing process. Quality control system enables the engagement of different techniques and process to ensure quality production output. Some of the techniques according to [4–6] has yielded tremendous results in the past and still remain relevant in the scope of industrial production and manufacturing till date.

However, some of the techniques and tools are gradually becoming obsolete and yielding reduced performance in term of output, therefore there is a need for gradual replacement of old methods with automation techniques, in order to sustain the tempo of productivity. This fact necessitate research in the area of quality assurance in industrial manufacturing. Adventure for development of new quality assurance system and methods of production that is automation based lead to evolution of Industry 4.0, which has since then changed the industrial manufacturing game [6]. Introduction of industry 4.0 with artificial intelligence into the manufacturing system has brought about replacement of old mechanical based method with new and smarter machine technique with automated system that uses sensors, this according to expert has changed procedures often used in quality checks in manufacturing sector. However, artificial intelligence has brought up application of robotics in industrial application and also the use of applications that has been empowered with sensors for automation capability. Smarter machine according to [5, 7] has led to enhanced productivity, improved quality standard and products.

Finally, in [5, 8] it was alluded that innovation of smarter machines, sensor enable machine has been a major addition from industry 4.0 technological disruption, that brought about intelligent manufacturing, it is on this premise that this study investigated system and process of adapting disruption technology in quality control monitoring to be able to achieve results oriented intelligent manufacturing.

1.1 Aim and objectives

1.2 Aim

The aim of the study is to carry out a study on application of disruptive application in managing quality system in intelligent manufacturing with a view to improving manufacturing process in organizations.

1.3 Objectives

There is a need to articulate objectives of the study, the objectives were synthesis from the gaps and emerging thoughts from literatures consulted. Therefore the following objectives are used in this study. They are to:

i. Investigate the state of disruption in quality monitoring in industrial manufacturing

ii. Examine the Drivers of Effective Quality Control System Monitoring in Intelligent Manufacturing

iii. Study Issues and Challenges Involved In Quality Control Systems in Intelligent Manufacturing

iv. Profile Critical Factors Influencing Adaptation of Effective Intelligent Manufacturing Process.

2. Literature review

In this section review of concept was carried out, and constructs were gleaned from the objectives and the aim and the title of this study. Therefore the review covers the following area: quality control, areas of disruption in quality monitoring and intelligent manufacturing. It includes the following: intelligent manufacturing system, quality control in manufacturing industry, industry 4.0 application disruption in quality assurance monitoring and challenges in quality control system.

2.1 Intelligent manufacturing system

The world of manufacturing environment has changed drastically in recent times, on account of industrialization. Also paradigm in production process, design and implementation has shifted in the direction of application of new generation applications, the applications could be found in design, monitoring and marketing industries. The new application has capability to accommodate high volume product processing, complex system and flexible schedule and sequence. The new system is referred to as intelligent system. The word intelligence comes from the new packages that comes with electronic tools that are now in popular use in industrial manufacturing [9]. The authors described intelligent system as electronic and automation replacement of traditional mechanical functions with new applications that uses sensors and sensitive Nano-tubes applications. In another clime, it is referred to as automatic system which found integration in design and monitoring system as pointed out in [9, 10].

Intelligent system is highly used in monitoring process during industrial manufacturing of products. Intelligent production systems are operated as a calibrated design and monitoring system, they are used in sequential monitoring of production system. They are used to monitor highly complex manufacturing system in order to achieve flexible manufacturing for instance, [10–12] submitted that intelligent system has enabled the use of sensor, optimization parameters, laser beam application and laser beam machine process. This according to [10] described the introduction of intelligent manufacturing system as a step ahead of traditional manufacturing system. The claim lies in the fact that intelligent manufacturing has capability of self-analyzing, self-learning, complexity apprehensions and large data storage. The self -analyzing attribute of systems of intelligent component is rooted in attribute of intelligent application that consist of the use of sensors. The sensors are mounted on tools and machine for quality control attribute in areas such as sequencing, intelligent scheduling, intelligent control and maintenance.

Moreover, in intelligent manufacturing, scheduling and sequencing are "sin-qua-non" there is interrelationship between the two concepts. The concepts tend to reoccurs throughout the production process because of its high utility. Also, [13] posited that intelligent algorithm was designed to assist in parametric calibration of some intelligent applications used in manufacturing industries. Algorithms are used in Asia and European industrial manufacturing sectors in permutation control of scheduling and sequencing operation. Scheduling operations involved scheduling of machine operations, allocation of resources of money, man, human and machine among others. [12, 14] in scheduling operations, parameters setting is of utmost importance. Some of the parameters was discovered by Johnson in 1965, scheduling was classified by Johnson into two categories; the flow shop and job shop. In concurrent intelligent scheduling, the flow shop utilizes Toyota production system in a way that the system could accommodate large volume of work per time, this view was supported in [14–17].

The choice of scheduling technique often depends on complexity, desired output and volume of system at hand. There are two conditions under which scheduling could be applied, the flow shop and job shop. Flow Shop in manufacturing and production process refers to high volume system that uses highly standardized equipment to ensure continuous flow of standardized products e.g. refineries, cement company, drinks production. Similarly, Job Shop is a low volume system, which periodically shift from one job to another. The production is often according to consumers' specification and orders are in small units [16, 17]. Moreover, sequencing is about methodical approach to processing loading jobs at work station. It describes the order in which jobs are processed or should be processed at work centers. In traditional scheduling operation the following rules subsists; the widely acceptable rules for scheduling operation in intelligent manufacturing according to [16–18] includes the following: First come First Serve (FCFS): Processing job in the order of arrivals at work center. Shortest Processing Time (SPT): Job are processed based base on length of processing time and Earliest Due Date (EDD): This rule sequences jobs according to their due date. Shortest due date are processed.

2.2 Quality control in manufacturing industry

In manufacturing sectors there have been gaps in the users request for product and the quantity of product manufactured to meet the requests, this trend, tend to permit packaging low quality products into markets. However, high level of quality is demanded as part of users demand expectation. In contemporary times, users expect faultless product, which has put more pressure on the manufacturer and their production outfits. Therefore, there should be shift from traditional ways of enforcing quality so as to enhance manufacturing productivity. Manufacturers' bid to improve quality has led cutting edge researches in quality control process, this therefore lead to new realm where quality control methods and processes are digitalized and this concept is termed quality 4.0. Quality 4.0 borders about the use of enhanced system to collect information about the design, behavior, use and performance of products in market. The quality 4.0 involve the use of benefits of Artificial intelligence (AI). Artificial intelligence works according to [19] through identifying faults in product with AI algorithm, the AI algorithm having been calibrated in fault identification would notify quality team of emerging faulty product. Similarly, quality control charts are often used alongside with AI in quality monitoring and also through pictorial illustration of quality trend in production batch, for instance in [20] simple statistical analysis method was used in interpreting quality results. It was noted in [21], that Japan adopted statistical method and tools in quality management. In Japan, certain quality control system has been in use in the manufacturing sector, for instance [20] submitted that the following system has been consistently engaged in intelligent manufacturing quality control in Japanese production process; Ishikawa, flow chart, figure and diagrams, pareto analysis, checklist and correlation among others. [21] identified six sigma template as being a useful tools as well in quality control monitoring process. Similarly, product reliability is also another measure introduced in ensuring quality in product development [22] presented quality control methods that could be used to enhanced productivity and safety [23]. Worked on systematic ways of presenting reliability analysis of control system, that, it has tendency of improving product reliability, strategies and implementation. System and reliability analysis is for the purpose of quality control during product design and manufacturing and control stage. The system was described in [22] as consist of mathematical model that can be used to describe relationship between product quality and control method. Product reliability process design include process conception, identification of product manufacturing process and robust design [22, 23].

2.3 Industry 4.0 application disruption in quality assurance monitoring

Industrial application in product manufacturing has introduced dimension of automation at various aspect of manufacturing. Generation of traditional manufacturing has metamorphosized into conventional system, thus, the new generation of industrial application covered different systems of conventional technology. In [24], systems that could enable product supply system to be optimized using key technologies such as cyber system, IoT (Internet of things), data analytics, big data, and expanded database. The key production points could be interlinked and interconnected or networked for effectiveness. Industry 4.0 enables digital interconnectivity among machine and work stations of production system. Similarly, industry 4.0 facilitates the interoperability of man and machine interconnected with digital guides [25]. However, industry 4.0 has different adaptability in different continent both in literary presentations and industrial application. In Japan, industry 4.0 is embellished as "New strategy", China captured it as Made-in-Chine 2025 as contained in {48} while United states according to [24] referred to industry 4.0 as "reindustrialization".

Industry 4.0 involved is centered on industrial reformation or reindustrialization and information applications contained in [26], this fact was corroborated by [27], which makes availability of quality supervisors and production managers. In nutshell, the component of industry 4.0 application in the portfolio of industrial manufacturing and production according as an inclusive component as viewed by [25, 28–30] and it consist of the following aspects in industrial production: IoT applications, learning factories, internet services, cloud computing and storage, cybernetics, cognitive computing, Artificial intelligence, robotics, data analytics, mechatronics, big data application, and sensor management among others.

2.4 Challenges in quality control system research (quality control implementation in

Right from the ancient days of product manufacturing, industrial application has experienced a lot of changes ranging from design parameters configuration of product design implementation. Companies and design expert has labor extensively to come up with perfect system, but there has always been one challenge or the other. Quality control system in intelligent manufacturing gas come with challenges. [31] Deloitte (2014) opined that, there are challenges associated with industry 4.0 digitalization and digital transformation.

Challenge of making reliable forecast is one of the major challenges often encounter in quality control management, however, [32] opined that the use of cyber technology equipment, physical system, artificial intelligence, big data would increase efficiency of running production system. Similarly, ineffective flow of materials and adequate planning has always been the bane of effective production system. [33, 34] argued that effective monitoring of flow system may not impart much on the production effectiveness unless match up with flow stream mapping.

Moreover, managing logistic tasks with material handling machine was suggested by [33, 35] that there is negative consequence that could come up on account of in appropriate adoption of material handling machine. Some of the consequences include absence: absence of route temperature, fluctuating environmental temperature, half-life of machine parts, and inadequate level of machine parts' management.

Furthermore, inadequate manpower planning, inadequate system calibration and continuous data transfer are some of the challenges that should be surmounted for effective quality control system during intelligent manufacturing [22, 36]. Finally, major challenge with the use of Artificial intelligence (AI) in creating intelligent manufacturing in recent time is the one that influence negatively, the variable of decision making and data inter- operability. [32] described parameters that influences AI adoption in control system to include: data parsimony, data transfer, data processing and manipulation, process speed, system efficiency and interoperability. Similarly, high data affinity or fidelity expectation, smart storage capacity and smart maintenance are components of smart and intelligent manufacturing that should be managed.

3. Materials and methods

In the context of this study, primary data was engaged and was collected from sampled Production managers, Production supervisors, Quality control officers and Information communication officers that are on ground at the selected locations of the research while Survey materials adopted structured questionnaire design in a closed structure manner as carried out in similar previous studies such as [13, 14, 16].

3.1 Material and tools

In this study different materials and tools were used, part of the materials used are A-4 papers for questionnaire production, Google forms, Google spread sheet, markers, pencils and biro. Analytical package of Statistical tools of SPSS was engaged in the processing of data collated from the respondents. Some of the tools include Relative Agreement Index (RAI), Mann–Whitney U-Test, Pearsons's Chi-square test and Student's T-test.

The Relative Agreement Index was calculated using the following relation.

$$\mathrm{RAI} = \sum \underline{\mathrm{Wi}} \qquad (1)$$

$$\mathrm{A} \times \mathrm{N}$$

where RAI = Relative Agreement Index, Wi = Weighted Sum, A = The number of items on Likert scale of 1–5.

N = individual weight of the scale item on Likert scale 1–5. The component of the Likert Scale include (SA: Strongly Agree (5), A: Agree (4); SD: Strongly Dis-agree (2); D:Dis-agree (1); N:Neutral (3)).

Survey design method was used in the study with population comprised of 100 manufacturing company both small and large scale. Similarly, sectionalized category of client were profiled and censored for data collection purpose, they include: Production Manager-PM; Quality Control Officer-QCO, Production Supervisor-PS; ICT Officer-Information Communication Technology Officer, the category of respondents were used in the study, they include the managers, officers and supervisors engaged in companies that are involved in consumer good products manufacturing, located at the at Federal Capital Territory in Abuja and Lagos state Nigeria. The questionnaires were initially pretested on five (5) respondents for content validation, their observation were later used to recalibrate the questionnaire into final form used for the analysis the view was supported in [13–15].

Summarily, the unique group of sample used in the study include sample size of seventy three (73) respondents that cut across the cadre of managers and supervisors in product manufacturing companies, i.e. Production Manager-PM; Quality Control Officer-QCO, Production Supervisor-PS; ICT Officer-Information Communication Technology Officer.

3.2 Questionnaire design

The data collection instrument adopted is a structured questions designed in Likert scale format of semantic rating scale 1–5. The questionnaire was designed in a way that allow for easy collation of data. It was divided into five sections. Section 1 was centered on the Bio-data information of the respondents, Section 2 was about categories of production manager, Section 3 was on Investigating the state of disruption in quality monitoring in industrial manufacturing, Section 4 was about the drivers of effective quality control system monitoring in intelligent manufacturing. Section 5 was centred on issues and challenges involved in quality control systems in Intelligent manufacturing while Section 6 focused on critical factors that influences an effective intelligent manufacturing system.

3.3 Operationalization of research variables

Question, analytical method	Scale	Variables	Reference
Q1–5 Respondents Bio-data information, Work experience. Analytical Methods: Descriptive Statistics, Percentage. Spearman Ranking.	Ordinal, Numeric, Likert Scale	Professional cadre, Gender, Types of managers, Qualification and cadre of managers on intelligent manufacturing system	[21]
Q6–14 Investigate the state of disruption in quality monitoring in industrial manufacturing Effectiveness of control systems Satisfaction level. Analytical method: Pearson's Chi-square, Kendal Tau teat and Relative Agreement Index and Spearman Ranking	Numeric Likert scale	Perception on state of adoption of quality control system for intelligent manufacturing	[20, 23]
Q15–23 Examine the Drivers of Effective Quality Control System Monitoring in Intelligent Manufacturing Analytical Method: Pearson's Chi-square, Relative Importance Index, Cronbach Alpha test, Man-Whitney U Test.	Numeric Likert scale	Drivers of Effective Quality Control System Monitoring in Intelligent Manufacturing system.	[21]
Q24–32 Study Issues and Challenges Involved In Quality Control Systems in Intelligent Manufacturing Analytical Methods: Pearson's Chi-square, Relative	Numeric Likert scale	Issues and Challenges Involved In Quality Control Systems in Intelligent Manufacturing	[20, 21]

Question, analytical method	Scale	Variables	Reference
Importance Index, Cronbach Alpha test, Man-Whitney U Test.			
Q33–42 Profile Critical Factors Influencing Effective Intelligent Manufacturing Analytical methods: Pearson's Chi-square, Kendal Tau teat and Relative Satisfaction Index and Spearman Ranking.	Numeric Likert Scale	Censoring satisfaction level on Critical Factors Influencing Effective Intelligent Manufacturing system	[20, 24]
Q43–51 Adaptable Areas of Disruptions in Quality Assurance Monitoring for Intelligent Manufacturing Analytical method: Relative Effectiveness Index, Kendal Tau Test. Spearman Ranking.	Numeric Likert Scale	Adaptable Areas of Disruptions in Quality Assurance Monitoring for Intelligent Manufacturing	[20]

3.4 Processing and distribution of questionnaire

In processing and administration of the data collection, survey design method was used while purposive sampling technique was adopted to pull together the respondents, as mentioned earlier in Section 2.2, the respondents used for this study are the production personnel, a total of eighty (80) questionnaire designed in Likert scale was sent to one eighty(80) managers and supervisors and officers that constitute respondents among which the completed and valid seventy three (73) questionnaire were used, this is similar to method adopted in studies of [13–15].

4. Results and discussion

4.1 Category of respondents

In the context of this study, different category of respondents were engaged in the study (**Table 1**). Twenty three (23) production supervisor are engaged, twenty (20) production managers, twenty (20) quality control officer and ten (10) ICT officers. All respondents are directly involved in production process of their various company at the research location.

4.2 Qualification of respondents

Manufacturing education background matters in the context of this research, level of enlighten would enable a respondent to contribute adequately towards valid response to the administered questionnaire or interview, Therefore, respondents qualification in this study is as presented in **Table 2**. It is obvious that the managers, supervisors and officers has more than one qualification. Among the 20 production managers sampled, 10 possessed trade certificates, 5 possessed masters and higher national diploma degree, this trend cut across the QCO, PS and ICTOs. The ICTOs has predominantly masters and bachelor degrees beside other certifications that are not classified in this category. Most of the production managers and supervisors has adequate practical background which enables them to be suitable for the purpose of the research.

Respondent cadre	Frequency	Percentage (%)
Production Supervisor	23	31.50
Production Manager	20	27.40
Quality Control Officer	20	27.40
ICT Officer	10	13.70
Total	73	100

Table 1.
Respondents Category.

Year experience	PM	QCO	PS	ICTO
BSc.[Bachelor]	—	5	5	3
MSc.[Masters]	5	5	10	7
HND[Diploma]	5	5	4	—
Trade Certificates	10	5	4	—
Total	20	20	23	10

Legend: Production Manager-PM; Quality Control Officer-QCO, Production Supervisor-PS; ICT Officer-Information Communication Technology Officer.

Table 2.
Respondents Qualification Manufacturing Experience.

4.3 Adaptable areas of disruptions in the quality assurance monitoring and calibration in production process

In **Table 3**, adaptable areas of disruptions in quality assurance monitoring was presented highlighting the responses of respondents. There are a lot of areas in intelligent manufacturing that disruptions has taken place. Some of the areas includes system design, implementation and monitoring. Areas covered in the study, 10 areas of disruptions was identified they include; process design, analytical technology, platform technology, operation technology, intelligent spindle system, sensor based control units, intelligent system powered psychometric system, 3d

Adaptable areas in intelligent manufacturing	RAI	Rank
Process Design	0.869	1st
Sensor based control units	0.786	2nd
Platform technology	0.771	3rd
Analytical technology	0.754	4th
Operation technology	0.732	5th
Intelligent Sequencing System	0.679	6th
Intelligent system powered psychometric system	0.657	7th
Intelligent spindle system	0.631	8th
3D Design System	0.579	9th

Legend: RAI—Relative Agreement Index.

Table 3.
Adaptable Areas of Disruptions in Quality Assurance Monitoring for Intelligent Manufacturing.

design system and intelligent sequencing system. This Process design is ranked 1st with RAI of 0.869; analytical technology is ranked 2nd with RAI value 0.786, platform technology with RAI 0.771 is ranked 3rd, operation technology is ranked 4th with RAI value of 0.754, intelligent spindle system also ranked 5th with RAI 0.732, sensor based control units ranked 6th, while intelligent system powered psychometric system and 3D design system and intelligent sequencing system were ranked 8th and 9th respectively with RAI values 0.631 and 0.579. In disruption that relates to intelligent manufacturing, there is always benefits often associated with paradigm shift in innovation technology. [36] opined that there are a lot of opportunities accrued to disruption that are beneficial although there is always negative resistance usually connected with a new change. There are risks associated with disruptions on account of new innovations often involved with the change.

However, any visionary organization should learn how to strategize in situation of associated risks mitigation. Intelligent manufacturing disruption has impacted strategic aspect of intelligent manufacturing that is changing the trend of games in product manufacturing. For instance, the following areas of intelligent manufacturing has been duly impacted; economy of product, economy of value chain, demand for product and changing faces of product as supported in [36, 37]. The associated risk in the disruption in intelligent manufacturing was modelled by [37–39] for monitoring of systemic sequencing of operation in product manufacturing. Disruption risk and pattern was modelled in [37] with Poisson ump process using random multiple system. Similarly parameters were sequenced and stochastically correlated to monitor quality diffusion process. However, internet of things (IoT), cyber-physical system, logistic 4.0, manufacturing 4.0, and hospital 4.0 are some of the aspect of disruption in intelligent manufacturing that formed the embodiment of application that has enhanced product production and manufacturing.

4.4 Drivers of effective quality control system monitoring in intelligent manufacturing

Drivers of effective quality control system culture adoption in intelligent manufacturing is presented in **Table 4**. From the data spread presented in the table, Effective monitoring of flow system with Relative Agreement Index (RAI) value of 0.873 is ranked 1st among other factors rated, Updated fault diagnostics and

Monitoring quality control parameters	RAI	Rank
Effective monitoring of flow system	0.873	1st
Updated fault diagnostics and rectification system	0.771	2nd
Effective design and production calibration system	0.769	3rd
Effective data extraction and transfer	0.754	5th
Intelligent cost control system	0.754	4th
Updated feedback control mechanism	0.732	6th
Adequate man-power planning	0.653	7th
Adequate machine part management	0.623	8th
Half live of machine parts	0.541	9th

RAI—Relative Agreement Index.

Table 4.
Drivers of Effective Quality Control System Culture Adoption in Intelligent Manufacturing.

rectification system RAI 0.771 is ranked 2nd, Effective design and production calibration system with RAI 0.769 ranked 3rd, Intelligent cost control system with RAI 0.754 ranked 4th, while Effective data extraction and transfer with RAI value of 0.754 is ranked 5th. However, Adequate machine part management with RAI 0.623 and Half live of machine parts with RAI value of 0.541 were ranked least with 8th and 9th position respectively.

Developing quality control is one of the important phase of manufacturing process. Quality control of product is very significant to the users and manufacturers. Quality to the manufacturer borders about correct design, authentic system calibration. Quality to consumer refers to manufactured product meeting users demand and expectation, therefore subjective in nature. ENQA (2009)in an adaptive study on standard and guidelines for quality assurance posited that there should be common reference point for quality assurance, keeping of quality register, procedure for quality assurance, keeping of quality register, documentation of procedure for the recognition of quality qualification and exchange of view among quality personnel. This could be encapsulated as a quality culture which should be keenly observed. Quality culture should be maintained so as to create a healthy production environment which guarantees customers production satisfaction. [40, 41] submitted further that quality culture would fine tune quality control value in a product system and this will lead to the reduction of political and geographical barrier). There are parameters that controls production in manufacturing industries, some of them includes: specification marketing, scheme and specification from purchasing units and agencies. Moreover, quality calibration at the outset of a manufacturing process usually ensures that quality product emerge from production system. Also, adoption of simple inspection table, and adoption of basic principles of sampling through sampling lot are cutting edge techniques that facilitates adoption of quality culture in product production, this toes the line of submissions in [20, 42].

4.5 Issues and challenges involved in adapting quality control systems in intelligent manufacturing

In this section challenges involved in the adapting of disruption application in intelligent manufacturing is outlined in **Table 5**. There are methods that could be used to control quality during manufacturing process [20, 21, 43] alluded that discrepancies in the document and. Process involved in manufacturing could be

Issues and challenges	PM	Rank	QCO	Rank	PS	Rank	ICTO	Rank
Machine–machine interaction	0.865	2nd	0.852	2nd	0.785	1st	0.773	1st
Man–machine interaction	0.873	1st	0.869	1st	0.721	2nd	0.653	2nd
Data quality	0.762	4th	0.754	4th	0.720	3rd	0.589	3rd
Cyber-security	0.771	3rd	0.771	3rd	0.698	4th	0.554	5th
Spare part management	0.762	4th	0.732	5th	0.657	5th	0.467	6th
Data Acquisition/Storage	0.657	5th	0.631	6th	0.589	6th	0.573	4th
Training Challenges	0.634	6th	0.543	7th	0.568	7th	0.457	7th
Testing cost & complexity	0.537	7th	0.557	7th	0.431	8th	0.456	8th

Legend: Production Manager-PM; Quality Control Officer-QCO, Production Supervisor-PS; ICT Officer-Information Communication Technology Officer.

Table 5.
Challenges involved in Adapting Quality Control System.

identified and process with different method, difficulty in the choice of correct method has been a challenge from time. In the context of this study, some significant challenges were profiled and processed. In relation to **Table 5**, some of the challenges profiled are stated as machine–machine interaction; man–machine interaction; data quality; cyber-security; spare part management; data acquisition/storage; training challenges and testing cost and complexity.

The respondents' opinion sample in this context about the challenges include that of production manager, quality control officer, production supervisors and ICT officers. The challenges are ranked in the appropriate order; machine–machine interaction, man–machine interaction, data quality, cyber-security, spare part management, data acquisition/storage, training challenges and testing cost & complexity. The production managers. ICT officers and Production supervisors ranked challenge of machine–machine interaction and man–machine interaction as 1st of the challenges; similarly, the same category of the managers ranked the same parameters 2nd according to the previous order. PS and ICO ranked data security and cyber security 3rd among other factors. Data quality was ranked 4th alongside Cyber security and data quality while data acquisition/storage; training challenges and testing cost and complexity were ranked 5th, 6th and 7th respectively by all categories of managers. Quality tools like Pareto diagram, Ishikawa diagram, Histogram, check-list, Flow chart and Control charts. However, statistical quality control method has been an age long method of quality control and this could be carried out with the aid of sample analysis, application of control charts and adoption of corrective measures.

4.6 Statistical test of level agreement of responses on issues and challenges involved in adapting quality control systems

Man Whitney U Test was carried out on the data in **Table 5** at significant level 0.05 two tailed, while the Test statistics results was presented in **Table 6**. The U-value is 24, the critical value of U at $P < 0.05$ is 8, therefore, the result is not significant at $P < 0.05$. Similarly, Z-score is 0.00. The P-value is 1.5 and 2.0, the result is not significant at 0.005. However, there is no significant difference in the opinion expressed by the respondents on the agreement level on Responses on issues and challenges involved in adapting quality control systems in intelligent manufacturing. It indicates high percentage of respondents in support of the fact that the KPI identified is in high order as presented on Likert scale 4 and 5 [23, 26].

From **Table 7** above, the Pearson Chi square results of the variables on Issues and Challenges Involved in Adapting Quality Control Systems is presented. From the table, the Pearson Chi-square row indicated that Chi value is 0.261, while P value is 0.05. This indicates that there is no statistical significance difference in association of respondents' opinion in the ranking of the issues and challenges involved in adopting control systems in intelligent manufacturing. The Null hypothesis is therefore accepted. It indicated that the factors were unanimously agreed upon. The implication lies in the fact that the listed factors in the tables were agreed upon by the respondents as critical challenge that poses threat and challenges to the intelligent manufacturing.

4.7 ANOVA of satisfaction level of facility managers on intelligent building systems' performance

Pearson's T and Student T test was carried out on mean of responses collated from the post occupation managers of the intelligent building used for the study and results presented in **Table 8**. The test was carried out to compare the difference in

DOI: http://dx.doi.org/10.5772/intechopen.93979

Officers/managers	Production manager		Quality control officer		Production supervisor		ICT officer		Std. deviation	Std. error
Statistic items	**N**	**Mean**	**N**	**Mean**	**N**	**Mean**	**N**	**Mean**	—	—
Machine–Machine interaction	1	1.500	1	1.500	1	1.500	1	1.500	—	—
	2	1.500	2	1.500	2	1.500	2	1.500	—	—
Man/machine Tradition	1	1.500	1	1.500	1	1.500	1	1.500	—	—
	2	1.500	2	1.500	2	1.500	2	1.500	—	—
Data Quality	1	1.500	1	1.500	1	1.500	1	1.500	—	—
	2	1.500	2	1.500	2	1.500	2	1.500	—	—
Cyber Security	1	1.500	1	1.500	1	1.500	1	1.500	—	—
	2	1.500	2	1.500	2	1.500	2	1.500	—	—
Spare parts Management	1	2.000	1	2.000	1	2.000	1	2.000	—	—
	2	1.000	2	1.000	2	1.000	2	1.000	—	—
Data Acquisition	1	2.000	1	2.000	1	2.000	1	2.000	—	—
	2	1.000	2	1.000	2	1.000	2	1.000	—	—
Training Challenges	1	2.000	1	2.000	1	2.000	1	2.000	—	—
	2	1.000	2	1.000	2	1.000	2	1.000	—	—
Testing and Complexity	1	1.500	1	1.500	1	1.500	1	1.500	—	—
	2	1.5000	2	1.5000	2	1.5000	2	1.5000	—	—

Table 6.
Man-Whitney U Student T-Test on Responses on Issues and Challenges Involved in Adapting Quality Control Systems.

Officers/managers	Production manager		Quality control office		Production supervisor		ICT officer	
Statistic items	**PS**	**KDT**	**PS**	**KDT**	**PS**	**KDT**	**PS**	**KDT**
Machine–Machine interaction	0.261	0.000	0.261	0.000	0.261	0.000	0.261	0.000
Man/machine Tradition	0.261	0.000	0.261	0.000	0.261	0.000	0.261	0.000
Data Quality	0.261	0.000	0.261	0.000	0.261	0.000	0.261	0.000
Cyber Security	0.238	0.030	0.238	0.030	0.238	0.000	0.238	0.000
Spare parts Management	0.238	0.164	0.238	0.164	0.238	0.091	0.238	0.091
Data Acquisition	0.261	0.000	0.261	0.000	0.261	0.000	0.261	0.000
Training Challenges	0.261	0.391	0.261	0.391	0.261	0.261	0.261	0.204
Testing and Complexity	0.261	0.000	0.261	0.000	0.261	0.261	0.261	0.000

Ho: There is no statistical difference in the managers' opinion as regards the Issues and Challenges Involved in Adapting Quality Control Systems.

Table 7.
Chi-Square Test on Issues and Challenges Involved in Adapting Quality Control Systems.

means of the post occupation facility managers and users. This is to check whether there variation among the group observed. The test results of equality of variance is as presented in the table. The test was performed at 95% confidence interval and all variables presented exhibited P-value higher than 0.05 i.e. $P > 0.05$. The result

ANOVA						
		Sum of squares	**df**	**Mean square**	**F**	**Sig.**
Machine to machine	Between Groups	1.000	3	.333		
	Within Groups	.000	0			
	Total	1.000	3			
Man to machine	Between Groups	1.000	3	.333		
	Within Groups	.000	0			
	Total	1.000	3			
Data quality	Between Groups	1.000	3	.333		
	Within Groups	.000	0			
	Total	1.000	3			
Cybersecurity	Between Groups	2.750	3	.917		
	Within Groups	.000	0			
	Total	2.750	3			
Spare parts	Between Groups	2.000	3	.667		
	Within Groups	.000	0			
	Total	2.000	3			
Data acquisition	Between Groups	2.750	3	.917		
	Within Groups	.000	0			
	Total	2.750	3			
Training challenges	Between Groups	.750	3	.250		
	Within Groups	.000	0			
	Total	.750	3			
Testing cost complexity	Between Groups	1.000	3	.333		
	Within Groups	.000	0			
	Total	1.000	3			

Table 8.
ANOVA of Satisfaction Level of Facility Managers on Intelligent Building Systems' Performance.

statistics implies that there is no significant difference in the means value the samples of the managers, therefore, the Null hypothesis is accepted, implying that there is no variation in the mean values of satisfaction level of production managers of the sampled companies. The reasons behind the difference could be linked to the background experience of the manager in intelligent manufacturing process.

Critical factors influencing intelligent manufacturing adaptation is presented in **Table 9**. The factors were broadly divided into four main categories, the technical related factors, political related factors, economic related factors and digital divide related factors. Main factors that are rated high belong to the technological related factors and were ranked 1st 2nd, 3rd and 4th, for instance, some production managers, quality control officers, production supervisor and ICT officers ranked category of factors that are technological related high, such as cost of talent, cost of technological failure, technological exchange problem and complexity in testing cost. Similarly, political related factors such as political will, technological content regulation and infrastructural environment provision are ranked 4th, 5th and 6th while digital divide factors were ranked 7th, 8th, 9th and 10th.

Critical factors parameters	PM	Rank	QCO	Rank	PS	Rank	ICTO	Rank
Technological Related Factor								
Cost of talent	0.885	1st	0.863	1st	0.785	1st	0.783	2nd
Cost of failure	0.883	2nd	0.861	2nd	0.721	5th	0.853	1st
Technological exchange	0.771	4th	0.757	5th	0.720	6th	0.759	4th
Testing cost and complexity	0.776	3rd	0.751	6th	0.698	7th	0.654	6th
Political Related Factor								
Government policy	0.766	7th	0.762	4th	0.657	9th	0.766	3rd
Technological content regulation	0.653	8th	0.651	7th	0.589	10th	0.651	7th
Provision of infrastructure environment	0.647	13th	0.643	8th	0.568	11th	0.457	14th
Political will for approval	0.653	8th	0.765	3rd	0.765	3rd	0.743	5th
Economic Related Factors								
Interplay of micro-economic variables	0.557	11th	0.587	9th	0.431	13th	0.456	14th
Interplay of macro-economic policy	0.766	7th	0.534	10th	0.678	8th	0.532	12th
Inter-continental technology transfer	0.765	5th	0.447	13th	0.787	2nd	0.521	13th
Digital divide related factors								
Data acquisition and storage	0.567	10th	0.531	12th	0.532	12th	0.654	8th
High regulation requirement	0.553	11th	0.534	10th	0.765	3rd	0.654	9th
Large state space	0.463	12th	0.503	11th	0.763	4th	0.503	11th

Legend: Production Manager-PM; Quality Control Officer-QCO, Production Supervisor-PS; ICT Officer-Information Communication Technology Officer.

Table 9.
Critical Factors Influencing Intelligent Manufacturing Adaptation.

Intelligent manufacturing is multidisciplinary in nature. This indicate that there are several intervening factors that influences success of intelligent manufacturing. Some of the significant factors are industrial chain factor, development factor, capital availability factor, political factor and socio-economic factors [44]. Submitted that chain factor, developing factor and capital factors influences the effectiveness of intelligent manufacturing system, the factors according to the study was described as engine to the effectiveness of intelligent manufacturing concept. Chain value system in supplying of materials for manufacturing is somehow significant, it borders about micro and macro supply chain system, and this toes the line of submissions in [45–47]. Similarly, there are issues that surrounds effective transition into intelligent manufacturing from traditional manufacturing system. While traditional manufacturing is transiting into intelligent manufacturing for the production of customized product involved principally. [46, 48] presented issues on important factors that should be taken into consideration in intelligent manufacturing.

The study revolves around actions to be taken to eliminate the challenges in manufacturing. Some of the actions include benchmarking capital factor, development factor and chain factor, the factors listed according to [48–51] formed the basis of the ideals in intelligent manufacturing and has tendency to transform traditional manufacturing production to customized product often associated with intelligent manufacturing. Moreover, [36] Adnan (2017) posited that manufacturing process and manufacturing organization responds to factors such as organization performance, organization culture, socio-economic reality and supply chain management. This fact is further corroborated in [52–55].

Officers/managers	Production manager		Quality control officer		Production supervisor		ICT officer		Std. deviation	Std. error
Statistic items	**N**	**Mean**	**N**	**Mean**	**N**	**Mean**	**N**	**Mean**	—	—
Technological Related Factor Cost of talent	1	1.300	1	1.300	1	**1.300**	**1**	**1.300**	—	—
	2	1.300	2	1.300	2	**1.300**	2	**1.300**	—	—
Cost of failure Technological exchange	1	1.300	1	1.300	1	**1.300**	**1**	**1.300**	—	—
	2	1.300	2	1.300	2	**1.300**	2	**1.300**	—	—
Testing cost and complexity Political Related Factor	1	1.300	1	1.300	1	**1.300**	**1**	**1.300**	—	—
	2	1.300	2	1.300	2	**1.300**	2	**1.300**	—	—
Government policy Technological content regulation	1	1.300	1	1.300	1	**1.300**	**1**	**1.300**	—	—
	2	1.300	2	1.300	2	**1.300**	2	**1.300**	—	—
Provision of infrastructure environment Political will for approval	1	2.000	1	2.000	1	**2.000**	**1**	**2.000**	—	—
	2	1.000	2	1.000	2	**1.000**	2	**1.000**	—	—
Economic Related Factors Interplay of micro-economic variables	1	2.000	1	2.000	1	**2.000**	**1**	**2.000**	—	—
	2	1.000	2	1.000	2	**1.000**	2	**1.000**	—	—
Interplay of macro-economic policy Inter-continental technology transfer	1	2.000	1	2.000	1	**2.000**	**1**	**2.000**	—	—
	2	1.000	2	1.000	2	**1.000**	2	**1.000**	—	—
Digital divide related factors Data acquisition and storage	1	1.300	1	1.300	1	**1.300**	**1**	**1.300**	—	—
	2	1.300	2	1.000	2	**1.000**	2	**1.300**	—	—
High regulation requirement	1	1.300	1	1.300	1	**1.300**	**1**	**1.300**	—	—
	2	1.300	2	1.000	2	**1.000**	2	**1.300**	—	—
Large state space	1	2.000	1	1.300	1	**2.000**	**1**	**2.000**	—	—
	2	1.000	2	1.000	2	**1.000**	2	**1.000**	—	—

Table 10.
Man-Whitney U Student T- Test on Critical Factors Influencing Intelligent Manufacturing Adaptation.

Man Whitney U Test was carried out on the data in **Table 9** and presented in **Table 10**, at significant level 0.05 two tailed. The U-value is 24, the critical value of U at $P < 0.05$ is 8, therefore, the result is not significant at $P < 0.05$. Similarly, Z-score is 0.00. The P-value is 1.5, the result is not significant at 0.005. However, there is no significant difference in the opinion expressed by the respondents on the agreement level by production managers, production supervisors, quality control officers, and ICT officers. it indicates high percentage of respondents in support of the critical factors influencing intelligent manufacturing adaptation which supports the fact that the KPI identified is in high order as presented on semantic rating Likert scale 5 and 4 [21, 22, 56].

5. Conclusions

The study has censored and profiled the issues, facts, ideals and factors that influences the adoption of disruptions in creating adaptive intelligent manufacturing with a view to creating enhanced productivity in intelligent manufacturing.

Literature review was conducted background for the study and set stage for identification of missing links and gaps. The Intelligent manufacturing has been noted to be fairly new in the developing countries [11–13]. Ten (10) areas of disruptions was identified in this study, include; process design, analytical technology, platform technology, operation technology, intelligent spindle system, sensor based control units, intelligent system powered psychometric system, 3D design system and intelligent sequencing system.

In the context of this study, some significant challenges were profiled and processed. Some of the challenges profiled include machine–machine interaction; man–machine interaction; data quality; cyber-security; spare part management; data acquisition/storage; training challenges and testing cost and complexity. The validity of the outcome of this research lies in applicability in expanding frontiers of knowledge in literary research, assistance to policy makers, assistance to the production managers and personnel among others. The study finally presents areas of disruptions in the quality assurance monitoring and calibration in production process, issues and challenges involved in quality control systems in manufacturing, emerging areas of application and recommendation for improvement.

Acknowledgements

The support of Building Informatics group members of Building Technology Department of College of Science and Technology and Center for Innovation and Research Development Covenant University is appreciated for various contributions and funding the OAPF.

Conflict of interest

The authors declare no conflict of interest in this research work. The funders had no role in the design of the study; in the collection, analyses, or interpretation of data; in the writing of the manuscript, or in the decision to publish the results.

Author details

Amusan Lekan[1,2]*, Clinton Aigbavboa[1], Moses Emetere[3] and James Owolabi[2]

1 Department of Construction Management and Quantity Surveying, Faculty of Engineering and Built Environment, University of Johannesburg, Doorfotein Campus Johannesburg, South Africa

2 Building Technology Department, College of Science and Technology, Covenant University, Ogun State, Nigeria

3 Department of Physics, College of Science and Technology, Covenant University, Ogun State, Nigeria

*Address all correspondence to: lekan.amusan@covenantuniversity.edu.ng

References

[1] Zaidin, N. H. M., Diah, M. N. M., Po, H. Y. & Sorooshian, S. (2018). Quality Management in Industry 4.0 Era. Jou

[2] Kagermann, H., Wahlster, W., Helbig, J. (2013). Recommendations For Implementing Thestrategic Initiative Industrie 4.0 - Final Report Of The Industrie 4.0 Working Group.

[3] Feigenbaum A.V (1991). Total Quality Control, New York: McGraw-Hill.

[4] Singh M, Khan IA, Grover S. Tools and Techniques for Quality Management in Manufacturing Industries. In: National Conference on Trends and Advances in Mechanical Engineering; 2012. pp. 853-859

[5] Hirsch-Kreinsen, H. (2014). Smart production systems. A new type of industrial process innovation. In Proceedings of the 2014 DRUID Society Conference, Copenhagen, Denmark. rnal of Management and Science 8(2), 82–91.

[6] Brüggemann H, Bremer P. Grundlagen Qualitäts management. Wiesbaden: Springer Vieweg; 2012

[7] Jaskó, Sz., Holczinger, T. and Skrop, A. (2019). Industry 4.0 laboratory of the University of Pannonia Nagykanizsa Campus. In: Science in Practice (XXXIV. Kandó Conference(2018) pp. 110–119. ISBN 978–963–449-096-8.

[8] Colledani M, Tolio T, Fischer A, Iung B, Lanza G, Schmitt R, et al. Design and management of manufacturing systems for production quality. CIRP Annals – Manufacturing Technology. 2014;**63**:773-796

[9] Holubek R, Kostal P. The Intelligent Manufacturing Systems. Journal of Computational and Theoretical Nanoscience. 2012 https://www.researchgate.net/publication/260918138.1-10

[10] Steven Liang, Manik Rajora;, Xianli Liu, Caixu Yue, Pan Zou; and Lihui Wang (2017) Intelligent Manufacturing Systems: A Review. International Journal of Mechanical Engineering and Robotics Research Vol. 7, No. 3, May 2018

[11] Teixidor, D I. Ferrer, J. Ciurana, and T. Özel, "Optimization of process parameters for pulsed laser milling of micro-channels on AISI H13 tool steel," Robotics and Computer-Integrated Manufacturing, vol. 29, no. 1, pp. 209–218, 2013

[12] Mukherjee, R, D. Goswami, and S. Chakraborty, "Parametric optimization of Nd: YAG laser beam machining process using artificial bee colony algorithm," Journal of Industrial Engineering, 2013.

[13] Haq, A.N, Ramanan,T.R, K. S. Shashikant, and R. Sridharan, (2010) "A hybrid neural network–genetic algorithm approach for permutation flow shop scheduling," International Journal of Production Research, vol. 48, no. 14, pp. 4217–4231, 2010

[14] Mirsanei, H.S, Zandieh,M., Moayed, M.J and Khabbazi,M.R (2011) "A simulated annealing algorithm approach to hybrid flow shop scheduling with sequence-dependent setup times," Journal of Intelligent Manufacturing, vol. 22, no. 6, pp. 965–978, 2011

[15] Bourne D., and M.S. Fox, (1984), "Autonomous Manufacturing: Automating the Job-Shop", Computer, Vol 17 No 9, pp. 76–88.CrossRefGoogle Scholar

[16] Fox MS, Lowenfeld S, Kleinosky P. Techniques for Sensor-Based Diagnosis. In: Proceedings of the International Joint Conference on Artificial Intelligence, August 1983. Los Altos, CA: William Kaufman Inc. Google Scholar; 1983

[17] Tian, Y, D. Liu, D. Yuan, and K. Wang, "A discrete PSO for two stage assembly scheduling problem," The International Journal of Advanced IIIIIOPHGManufacturing Technology, vol. 66, no. 1–4, pp. 481–499, 2013.

[18] Adedayo , A, Olu, O,Obamiro, A. (2006) Operations Research. Second Edition University Press Ltd. 13-16.

[19] Sreelekha Panda (2018) Impact of AI in Manufacturing Industries. International Research Journal of Engineering and Technology (IRJET), 05 Issue: 11,1786

[20] Miroslav Prístavka*, Martina Kotorová, Radovan Savov (2016) Quality Control in Production Processes. Acta Technologica Agriculturae 3Nitra, Slovaca Universitas Agriculturae Nitriae, 2016, pp. 77–83

[21] Kučera, M. – Bujna, M. – Korenková, M. – Haas, P. (2014). Possibilities of using ecological fluid in agriculture. In Advanced Materials Research, vol. 1059, 2014, special iss., pp. 61–66.

[22] Beata Mrugalska*, Edwin Tytyk (2015) Quality control methods for product reliability and safety. 6th International Conference on Applied Human Factors and Ergonomics (AHFE 2015) and the Affiliated Conferences, AHFE 2015. Procedia Manufacturing 3 (2015) 2730–2737

[23] Levin TT. Kalal, Improving Product Reliability: Strategies and Implementation. Chichester: John Wiley & Sons Ltd; 2003

[24] Adrienn Skrop - Tibor Holczinger - Krisztián Bakon - Bálint Mihalics - Szilárd Jaskó(2017). Industry 4.0 - Challenges in Industrial Artificial Intelligence. Ii. International Scientific Conference on Tourism and Security, 4 December, 2018,12–25.

[25] Schmidt, R., Möhring, M., Härting, R.C., Reichstein, C., Neumaier, P., Jozinovi'c, P. (2015). Industry 4.0-potentials for creating smart products: Empirical research results. In International Conference on Business Information Systems; Springer: Cham, Germany,Volume 12, pp. 16–27.

[26] Lanza, G., Nyhuis, P,; Ansari, S.M., Kuprat, T., Liebrecht, C. (2016). Befähigungs-und Einführungsstrategien für Industrie 4.0. ZWF Zeitschrift Wirtschaftlichen Fabrikbetrieb,111, 76–79.

[27] Lichtblau K, Stich V, Bertenrath R, Blum M, Bleider M, Millack A, et al. Industrie 4.0 Readiness. Impuls-Stiftung des VDMA Aachen-Köln. 2015;**52**:1-77

[28] Munir S, Stankovic JA, Liang CJM, Lin S. Cyber Physical System Challenges for Human-in-the-Loop Control. In Proceedings of the Presented as part of the 8th International Workshop on Feedback Computing, USENIX, San Jose, CA, USA, 24–28 June 2013. 2013;**4**:1-4

[29] Gilchrist, A. (2016). Industry 4.0: The Industrial Internet of Things; Apress: Berkeley, CA, USA.

[30] Schumacher A, Erol S, Sihn W. A maturity model for assessing industry 4.0 readiness and maturity of manufacturing enterprises. Procedia CIRP. 2016;**52**:161-166

[31] Delloite (2016) Disruption in Manufacturing. American Psychological Association Publication. 10–25.

[32] Béla Illés1,a, Péter Tamás1,b, Péter Dobos1,c, Róbert Skapinyecz1,( ) New Challenges for Quality Assurance of Manufacturing Processes in Industry 4.0. Solid State Phenomena Vol. 261, 481–486

[33] Cselényi, B. Illés: PlaPing Su and Shuyuang Liu (2015) DualSourcing in Managing Operational and Disruption

Risks in Contract Manufacturing. InternationalJournalof Production Research. Vol. 53(1), 291–306.

[34] Gubán, A. Kása R.: Conceptualization of fluid flows of logistificated processes, Advanced logistic systems 7 pp 27–24

[35] Al-Tit AA. Factors affecting the organizational performance of manufacturing firms. International Journal of Engineering Business Management. SAGE. 2017;**9**:1-9 https://doi.org/10.1177/1847979017712628

[36] Arijit Mukherjee, Abhijit Chakraborty, Mr.Sujit Kumar Garai (2016) Essence of Quality Control in Small Manufacturing Industry. IRA-International Journal of Technology & Engineering. ISSN 2455–4480; Vol.03, Issue 03.

[37] Sua P, Liu S. Dual sourcing in managing operational and disruption risks in contract manufacturing. International Journal of Production Research, 2015. 2015;**53**(1):291-306 h ttp://dx.doi.org/10.1080/00207543.2014.957876

[38] Allon, G., and J. A. V. Mieghem. Forthcoming. "Global Dual Sourcing: Tailored Base-surge Allocation to Near and Offshore Production."Management Science 56 (1): 110–124

[39] Hou, J., A. Z. Zeng, and L. Zhao. 2010. "Coordination with a Backup Supplier through Buy-back Contract under Supply Disruption." Transportation Research Part E: Logistics and Transportation Review 46 (6): 881–895.

[40] ENQA (2019) Standards and Guidelines for Quality Assurance in Europe http://enquin.eu/oop-content/upload/2020/ES9.

[41] El Safty S. Five Essential Ingredients for a Quality Culture. PEX Network Annual Survey. 2020:1-10

[42] Sinha (2020) Quality Control (QC) Definition, Importance and Tools of Quality Control. http://www.yourarticlelibrary.com/product-management/quality-control-qe-definitionimportance-and-tools-of-quality-control

[43] Korenko, M. – Beloev, C. I. – Kaplík, P. (2013). Quality Control Using PPAP Method. Ruse : Angel Kanchev University of Ruse. (Scientific monograph) . 105–109.

[44] Liu Xinyuan, Cheng Qing and Guokun (2017) Influencing Factors of Intelligent Manufacturing: Empirical Analysis Based on SVR Model> Information Technology and Quantitative Management (ITQM) Procedia Computer Science. 122, 1024–1030.

[45] Guo Q, Zhang M. An agent-oriented approach to resolve scheduling optimization in intelligent manufacturing[J]. Robotics and Computer-Integrated Manufacturing. 2010;**26**(1):39-45

[46] Manu cloud[EB/OL].[2014–03 - 01] http: / /www. manucloud-project. eu/.

[47] Choy KL, Lee WB, Lau H, et al. Design of an intelligent supplier relationship management system for new product development[J]. International Journal of Computer Integrated Manufacturing. 2004;**17**(8): 692-715

[48] Wang Webin (2019) Analysis of the Factors Influencing the relevant development fromTraditional Manufacturing. Management and Education (ICSSME 2019),185–190.

[49] Bao Sasha. Promoting the Transformation from Traditional Manufacturing to Intelligent Manufacturing [N]. Jianghuai Times, 2018-11-13(003)

[50] Meng Fansheng, Zhao Gang. Research on Influencing Factors of

Traditional Manufacturing to Intelligent. Manufacturing [J]. Scientific and Technological Progress and Countermeasures, 2018.189

[51] Li, T., S. P. Sethi, and J. Zhang. 2013. "How does Pricing PowerAffect a Firms Sourcing Decisions from Unreliable Suppliers?" International Journal of Production Research 51 (23–24): 6990–7005

[52] Abdul Rasid S, Isa C, Ismail W. Management accounting systems, enterprise risk management and organizational performance in financial institutions. Asian Rev Accounting. 2014;**22**(2):128-144. DOI: 10.1108/ARA-03-2013-0022

[53] Felício J, Couto E, Caiado J. Human capital, social capital and organizational performance. Manage Decis. 2014; **52**(2):350-364. DOI: 10.1108/MD-04-2013-0260

[54] Fernández-Pérez V, García-Morales V. Bustinza-Sánchez, Ó (2012) The effects of CEOs' social networks on organizational performance through knowledge and strategic flexibility. Person Rev. 2012;**41**(6):777-812

[55] Wen Gan. The gunfire of the transformation of traditional intelligent manufacturing industry has sounded [J]. Internet Weekly, 2017.DOI:10.1108/00483481211263719.

[56] Bierma T, Waterstraat F. Overcoming Barriers to Pollution Prevention in Small Businesses, Waste Management and Research Center Reports RR-E075. Illinois Department of Natural Resources: Champaign, IL, USA; 2001

Chapter 5

Analysis and Two-Dimensional Modeling of Directional Coupler Based on Two Coplanar Lines

Anouar Acheghaf and Naima Amar Touhami

Abstract

This chapter is dedicated to physical modeling and numerical characterization of directional coupler based on two coplanar lines using the general theory of coupled lines. The modeling in this chapter is two-dimensional due to the chosen numerical method (MOMs), for that purpose the analysis is divided into steps, we started by analyzing and modeling a micro-coplanar line in the quasi-TEM approximation using Green's functions and the integral equation method then we conclude by using the telegraphist equations and the results of the first step to modeling a couple of micro-coplanar lines.

Keywords: micro-coplanar lines CPW, directional coupler, Green's functions, integral equation method, MOMs, quasi-TEM approximation, general theory of coupled lines

1. Introduction

It is well known that MICs are based on the use of the technology of planar circuits printed partially or entirely, on a flat surface of a dielectric, by an etching operation. The entire circuit can be produced in large numbers at low cost by photolithography.

The technical characteristics of MICs are their small size, their low weight, and their high reliability.

At the end of the 1970s, the advancement of planar circuit technology coupled with the rapid development of microwave semiconductor components, particularly MESFET on gallium arsenide and advances in manufacturing materials technologies, were at the origin of the emergence of microwave monolithic integrated circuit (MMIC) technology. In this technology, the passive and active circuits and their interconnections are produced in large numbers on the same substrate.

In MIC technology, the structure of the planar waveguide consists of block elements according to the development of various functional components or subsystems. The study of planar waveguide structures was an important research subject in the field of MIC circuits. in recent years, the explosive development of commercial microwave applications for the general public, has considerably increased research activities in this field on the one hand, to explore the various new configurations of planar circuits, on the other hand to precisely characterize their electrical performance.

Planar transmission lines are the most essential point in PCM circuits. in the late sixties, with the availability of dielectrics with high dielectric constants of low loss dielectric materials and with the increasing demand for miniaturized microwave circuits for space applications. The intensity of interest in micro-ribbon circuits was renewed. This resulted in the rapid development of the use of micro-stipe lines.

At that time, two other types of planar transmission lines were also invented: they are slotted lines, and coplanar (CPW) respectively proposed by S.B. Cohn and C.P. Wen.

With the growth in operating frequencies, particularly in the millimeter band, the use of the traditional microstrip line becomes problematic because of the increase in losses, the presence of higher order modes and parasitic couplings. In this regard, the interest in uniplanar waveguide structures, using only a one substrate face, was renewed.

Uniplanar transmission structures include the coplanar lines which is modeled in this chapter, the slotted lines, and the two-ribbons coupled lines. These structures have many advantages over microstrip lines, such as easy production of successive parallel connections of passive or active components without the need to resort to metallized holes to a ground plane, low dispersion [1, 2].

The physical modeling and then the numerical characterization of these planar transmission lines have been an important axis in the field of scientific research in recent years. Many techniques and numerical methods have been developed and used for the numerical characterization of uniplanar structures. In general, the numerical methods for the study of MMICs can be classify into two groups: the first includes the integral methods like MOMs, NEWTON-COTES FORMULAS, ... and the other group is derivative like the FDTD, TLM ... DF method.

The analysis in this chapter is devoted to modeling a directional coupler in CPW technology by modeling one and two micro-coplanar lines on a dielectric substrate, the problem starts with solving the poison's equation which is transformed to the integral equation in which the unknown is the charge density ρ on the metal strip, by using the method of the Green's function [3] and the conformal mapping [4] technique. Subsequently we apply the numerical method (MOMs) [5] to solve the integral equation for obtaining the unknown function which is used to determine the variation of capacitance C, the value of the characteristic impedance Zc, and the effective permittivity εeff.

The second part of this analysis is to search the expression of the characteristic impedance for the odd Z_{co} and even Z_{ce} modes as a function of the geometric dimensions of the directional coupler in CPW technology and the coupling coefficient K as a function of the gap ? between the two strip lines by using the general theory of coupled lines (telegraphist equations) and the results of the first part.

2. Statement of the problem in quasi-TEM mode

In this part we focused to the formulation of the problem studied by determining the different characteristics of one and two micro-coplanar lines in the quasi-TEM approximation, using the integral equation method using the Green's function technique and the conformal mapping. The integral equation method is suitable for planar structures and it's most used to solve the electromagnetic problems.

The problem starts with solving the poison's equation Eq. (1) to obtaining the linear charge density ρ on the central metal strip shown in **Figure 1**.

$$\left(\frac{\partial^2}{\partial x^2} + \frac{\partial^2}{\partial y^2}\right)\psi(x,y) = -\frac{1}{\varepsilon}\rho(x,y) \tag{1}$$

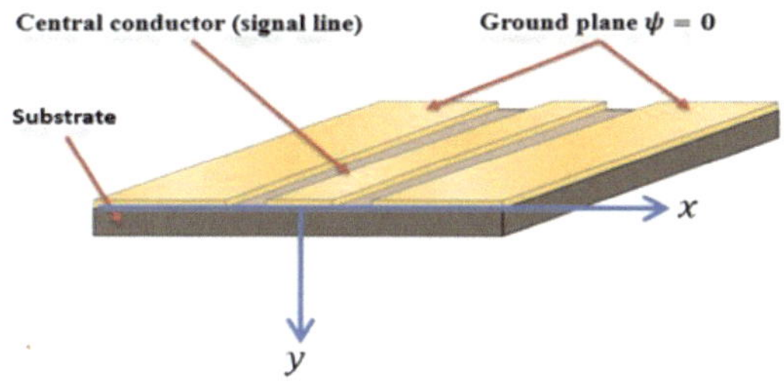

Figure 1.
Micro-coplanar line.

With $\rho(x,y) = \rho(x)$ for $y = 0$ and $x \in$ strip and $\rho(x,y) = 0$ elsewhere.

The determination of $\rho(x, y = 0)$ makes it possible to evaluate the capacity C per unit length of the coplanar line, and that will allow us to determine the other characteristic parameters.

To solve the poison's equation, we inverting the Laplacian operator using the Green's function to the integral operator Eq. (2) to form the integral equation:

$$\psi(x,y) = \frac{1}{\varepsilon} \int G(x, y|x_0, y_0) \rho(x) dl \tag{2}$$

The determination of the Green's function G corresponding to the studied problem constitute the most delicate step. Once this function is obtained, the second step is to solve numerically the integral equation by the method of MOMs.

3. Determination of the Green's function

This function used to form the integral equation Eq.(2), thereby it represents the inverse of Laplacian operator Eq.(3), where the point (x,y) is said field point created by a unit charge $(1C)$ at the point (x_0, y_0) said source point.

$$G_0(x, y|x_0, y_0) = \left(\frac{\partial^2}{\partial x^2} + \frac{\partial^2}{\partial y^2}\right)^{-1} \tag{3}$$

$$\left(\frac{\partial^2}{\partial x^2} + \frac{\partial^2}{\partial y^2}\right) G_0(x, y|x_0, y_0) = -\delta(x - x_0)(y - y_0) \tag{4}$$

With δ is the Dirac function, so the Green's function G_0 is written in the following form:

$$G_0(x, y|x_0, y_0) = -\frac{1}{2\pi} ln \sqrt{(x - x_0)^2 + (y - y_0)^2} + cte \tag{5}$$

To calculate the Green's function corresponds to the coplanar line without the central strip shown in **Figure 2a**, first we calculate the Green's function of the electrical potential created by a distribution of charges between two infinite ground planes shown in **Figure 2b** using the multiple image method [2], given by [3] as follow:

$$G_0(x, y|x_0, y_0) = -\frac{1}{4\pi} \sum_{n=-\infty}^{n=+\infty} ln \left(\frac{(y - y_0)^2 + (x - x_0 - 2na)^2}{(y - y_0)^2 + (x + x_0 - 2na)^2}\right) \tag{6}$$

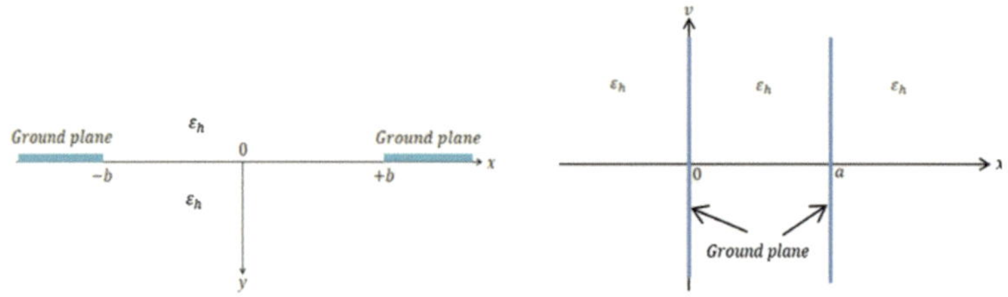

Figure 2.
(a) Micro-coplanar line without central conductor; (b) structure with two infinite ground planes.

With a is the gap between the two infinite metal plate **Figure 2a**, its sum given by [6] as:

$$G_0(x, y|x_0, y_0) = -\frac{1}{4\pi} \, ln \left(\frac{sin^2 \frac{\pi}{2a}(x - x_0) + sh^2 \frac{\pi}{2a}(y - y_0)}{sin^2 \frac{\pi}{2a}(x + x_0) + sh^2 \frac{\pi}{2a}(y - y_0)} \right) \quad (7)$$

The next step consists of applying a suitable conformal mapping which makes it possible to transform the Green's function into the given structure in **Figure 2a**, it's given as:

$$G(x, y|x_0, y_0) = -\frac{1}{4\pi} \, ln \left(\frac{sin^2\left(\frac{1}{2}\left(cos^{-1}\left(\frac{x}{b.z}\right) - cos^{-1}\left(\frac{x_0}{b.z_0}\right)\right)\right) + sh^2\left(\frac{1}{2}\left(ch^{-1}(z) - ch^{-1}(z_0)\right)\right)}{sin^2\left(\frac{1}{2}\left(cos^{-1}\left(\frac{x}{b.z}\right) + cos^{-1}\left(\frac{x_0}{b.z_0}\right)\right)\right) + sh^2\left(\frac{1}{2}\left(ch^{-1}(z) - ch^{-1}(z_0)\right)\right)} \right) \quad (8)$$

With: $z = \left(\frac{\alpha+\beta}{2}\right)^{1/2}$; $z_0 = \left(\frac{\alpha_0+\beta_0}{2}\right)^{1/2}$; and $\alpha = \left(\frac{x}{b}\right)^2 + \left(\frac{y}{b}\right)^2$; $\beta = \left(\alpha^2 - 4\left(\frac{x}{b}\right)^2\right)$.

The Green's function Eq. (8) is also a reciprocal function [7], and it's can be expressed as the superposition of two functions for non-homogeneous [8] middle, and for isotropic or anisotropic [9–10] middles.

So, for a non-homogeneous middle shown in the **Figure 3**, the Green's function is written as [3]:

$$G(x, y|x_0, y_0) = -\frac{1}{4\pi} \left(ln \left(\frac{sin^2(R^-) + sh^2(T^-)}{sin^2(R^+) + sh^2(T^-)} \right) + R. \, ln \left(\frac{sin^2(R^-) + sh^2(T^+)}{sin^2(R^+) + sh^2(T^+)} \right) \right) \quad (9)$$

With: $R^{\pm} = \frac{1}{2}\left(cos^{-1}\left(\frac{x}{b.z}\right) \pm cos^{-1}\left(\frac{x_0}{b.z_0}\right)\right)$; $T^{\pm} = \frac{1}{2}\left(ch^{-1}(z) \pm ch^{-1}(z_{0.})\right)$

And: $R = \frac{\varepsilon_1 - \varepsilon_2}{\varepsilon_1 + \varepsilon_2}$.

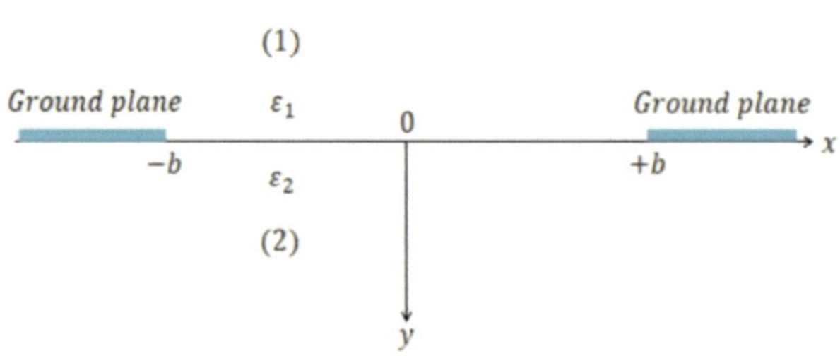

Figure 3.
Micro-coplanar (non-homogeneous middle).

4. Numerical characterization of the integral equation

The linear charge density ρ on the metal strip shown in **Figure 4** is related to the distribution potential ψ in the cross-section of the coplanar line, and it's zero elsewhere. For that reason, the integral equation Eq. (2) can be written as:

$$\psi(x,y) = \frac{1}{\varepsilon}\int_{-w/2}^{w/2} G(x,y|x_0,y_0).\rho(x_0,y_0)dl_0 \tag{10}$$

The central strip conductor is considered infinitely thin, that allows us to write:

$$\psi(x) = \frac{1}{\varepsilon}\int_{-w/2}^{w/2} G(x|x_0).\rho(x_0)dx_0 \tag{11}$$

Assuming that the central strip is submitted to a unit potential, so the equation Eq. (11) is convenient to solve numerically using the method of moments. This method is divided into two stages, firstly by developing $\rho(x_0,y_0)$ in series of N basic functions Eq. (12) in the form of rectangular pulses Eq. (13), secondly by using the Galerkin procedure which allows to write equation Eq. (11) as a linear equations system Eq. (14):

$$\rho_j(x_0) = \sum_{j=1}^{N} \alpha_j f_j(x_0) \tag{12}$$

$$f_j(x_0) = \begin{cases} 1 & \text{if } x \in \left[x_j - \frac{h}{2}; x_j + \frac{h}{2}\right] \\ 0 & \text{alsewhere} \end{cases} \tag{13}$$

$$\sum_{i=1}^{N} \psi_i(x_0) = \sum_{i=1}^{N}\sum_{j=1}^{N} \frac{\alpha_j}{\varepsilon_e}\int_{x_j-\frac{h}{2}}^{x_j+\frac{h}{2}} G(x_i|x_j)dx_0 \tag{14}$$

So, the linear equations system (14) can be written in the following matrix form:

$$[\alpha_j] = [M_{ij}]^{-1}.[\psi_j] \tag{15}$$

With $[M_{ij}]$ is the matrix of the Green's function obtained by the inversion of the linear equations system Eq. (14), defined on the central strip of the structure

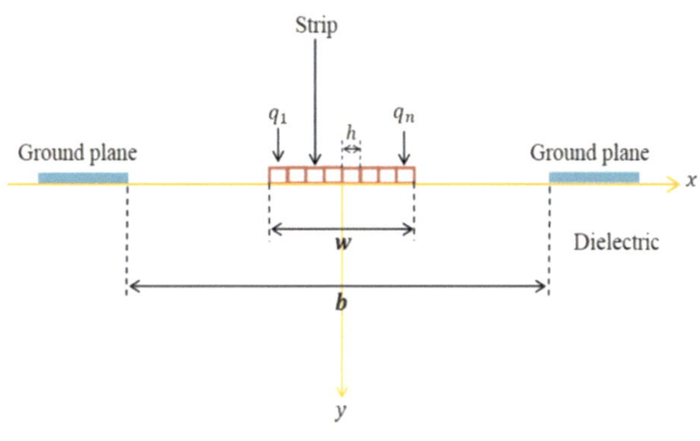

Figure 4.
Discretization of the central strip.

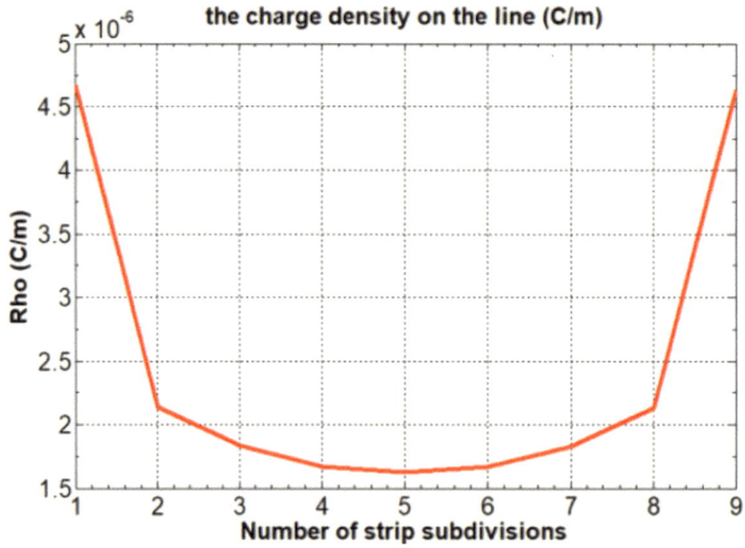

Figure 5.
Variation of the charge density on the metal strip.

studied show in **Figure 4**. And **Figure 5** shows the variation of the charge density $\rho_j(x_0)$ (solution of the integral equation) as a function of subdivisions N of the central strip of with w.

5. Characteristics of the coplanar line

In this part we present the numerical results of the variation of the characteristic impedance Z_c of the transmission line shown in **Figure 6** and of the variation of the effective dielectric permittivity ε_{eff} as a function of the ratio w/b shown in **Figure 7**. The line is assumed to be lossless $(R = G = 0)$, then its characteristic impedance is given by:

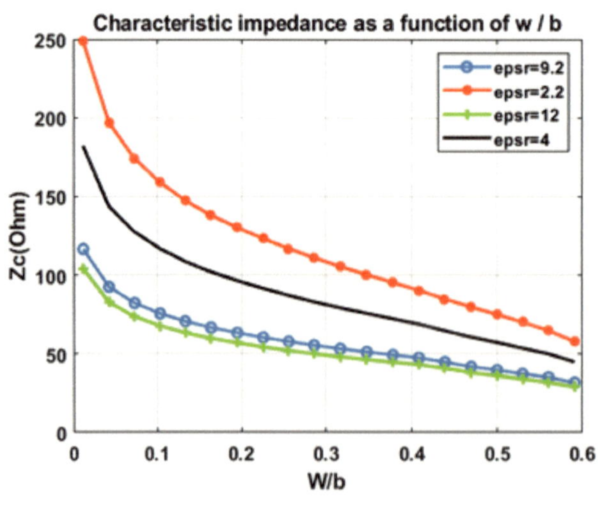

Figure 6.
Variations of the characteristic impedance for different ε_{eff}.

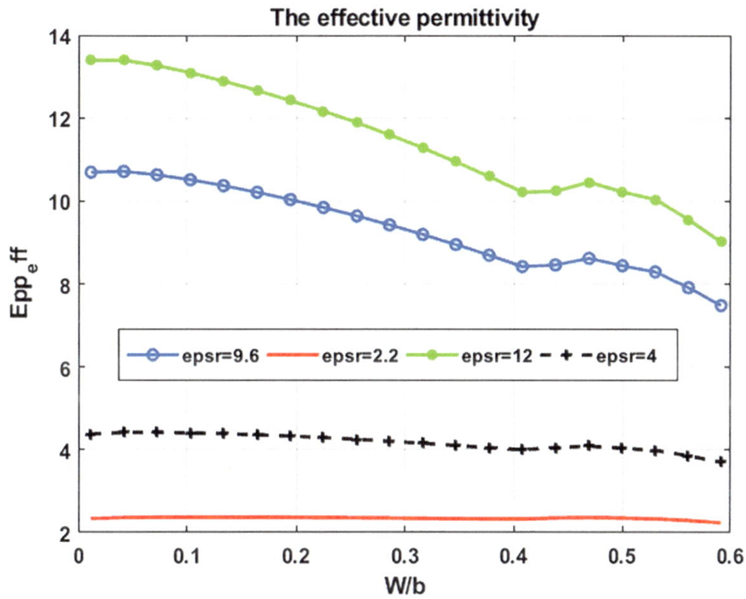

Figure 7.
Variations of the effective permittivity.

$$Z_c = \sqrt{\frac{L}{C}} = v_p L = \frac{1}{v_p C} \tag{16}$$

$$C = \frac{\sum_{j=1}^{N} \rho_j}{\psi} \tag{17}$$

$$L = \frac{1}{v_{pa}^2 C} = \frac{1}{\varepsilon_0 \mu_0 C} \tag{18}$$

$$\varepsilon_{eff} = \left(\frac{v_p}{v_{pa}}\right)^2 \tag{19}$$

With v_p is the propagation velocity in the coplanar line and v_{pa} is the propagation velocity in the air, while L and C is the is the inductance and capacity per unit length of the micro-coplanar line. And ψ is the unit potential.

6. Analysis of a directional coupler in CPW technology

A coupled coplanar line configuration consists of two transmission lines placed parallel to each other and in proximity as shown in **Figure 8**. In such a configuration there is a continuous coupling between the electromagnetic fields of the two lines. Coupled lines are utilized extensively as basic elements for coplanar directional coupler which is the subject of this study, filters, amplifiers, and a variety of other useful circuits.

Because of the coupling of electromagnetic fields, a pair of coupled lines can support two different modes of propagation. These modes have different characteristic impedances Z_{co} for odd mode, and Z_{ce} for even mode. The general theory of coupled lines (telegraphist equations) is used as method of analysis to determine those impedances for each mode of propagation, and to calculate the coupling coefficient K.

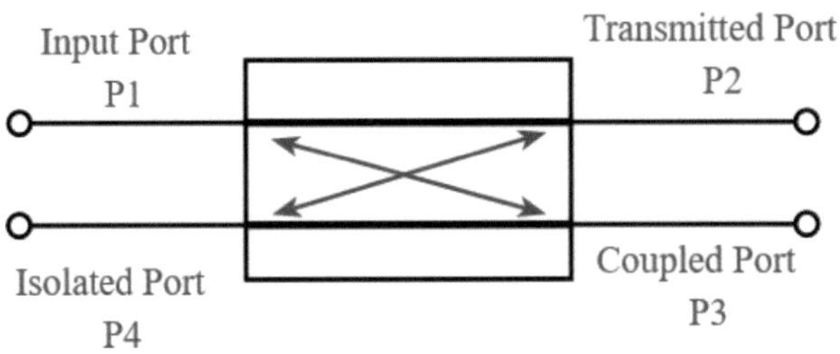

Figure 8.
Directional coupler.

We suppose that the propagation is along the axis OZ, the telegraphist equation is written [10]:

$$\begin{cases} -\frac{d}{dz}[V] = [Z].[I] \\ -\frac{d}{dz}[I] = [Y].[I] \end{cases} \tag{20}$$

With $[Z]$ is the impedance matrix and $[Y]$ represents the admittance matrix of the directional coupler. This system can be written as:

$$\begin{cases} -\frac{dv_1}{dz} = Z_1 i_1 + Z_m i_2 \\ -\frac{dv_2}{dz} = Z_m i_1 + Z_2 i_2 \\ -\frac{di_1}{dz} = Y_1 v_1 + Y_m v_2 \\ -\frac{di_2}{dz} = Y_m v_1 + Y_m v_2 \end{cases} \tag{21}$$

With Z_1, Z_2 are the impedances of the coupled lines per unit length, and Y_1, Y_2 their admittance, where Z_m, Y_m are the mutual impedance per unit length and the mutual admittance per unit length respectively.

The system Eq. (21) is a system of homogeneous first-order differential equations with constant coefficients. The resolution of this system gives the voltage and the current for even and odd propagated modes is as follow:

$$v_1 = \left(A_1 e^{-\gamma_e Z} + A_2 e^{-\gamma_e Z}\right) + \left(A_3 e^{-\gamma_o Z} + A_4 e^{-\gamma_o Z}\right) \tag{22}$$

$$v_2 = R_e\left(A_1 e^{-\gamma_e Z} + A_2 e^{-\gamma_e Z}\right) + R_o\left(A_3 e^{-\gamma_o Z} + A_4 e^{-\gamma_o Z}\right) \tag{23}$$

$$i_1 = Y_{e1}\left(A_1 e^{-\gamma_e Z} - A_2 e^{-\gamma_e Z}\right) + Y_{o1}\left(A_3 e^{-\gamma_o Z} - A_4 e^{-\gamma_o Z}\right) \tag{24}$$

$$i_2 = Y_{e2} R_e\left(A_1 e^{-\gamma_e Z} - A_2 e^{-\gamma_e Z}\right) + Y_{o2} R_o\left(A_3 e^{-\gamma_o Z} - A_4 e^{-\gamma_o Z}\right) \tag{25}$$

Where $R_{e,o}$ and $Y_{e1,2;o1,2}$ are functions depending on the impedance and admittance of the coupled line. As a result, the propagation constants of the two modes are expressed as a function of linear capacitances and inductances:

$$\gamma_{e,o} = j\beta_{e,o}$$

$$= j\frac{\omega}{\sqrt{2}}\sqrt{(L_1C_1 + L_2C_2 - 2L_mC_m) \pm \sqrt{(L_2C_2 - L_1C_1)^2 + 4(L_mC_1 - L_2C_m)(L_mC_2 - L_1C_m)}} \tag{26}$$

$$R_{e,o} = \frac{(L_2C_2 - L_1C_1) \pm \sqrt{(L_2C_2 - L_1C_1)^2 + 4(L_mC_2 - L_1C_m)(L_mC_1 - L_2C_m)}}{2(L_mC_2 - L_1C_m)} \tag{27}$$

Therefore, the characteristic impedances of the two coupled lines for even and odd mode are:

$$Z_{ce} = \frac{\omega}{\beta_e}\left(L_1 - \frac{L_m}{R_o}\right) = \frac{\beta_e}{\omega}\left(\frac{1}{C_1 - R_eC_m}\right) \tag{28}$$

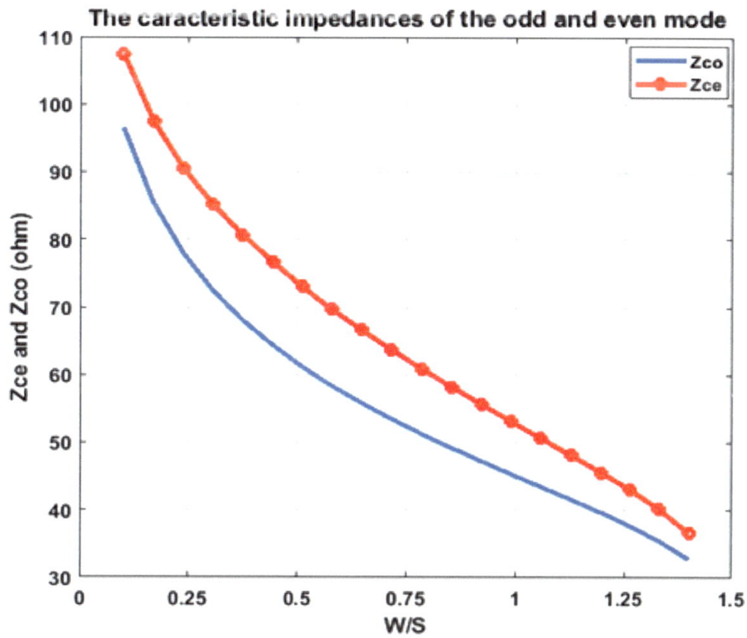

Figure 9.
Variations of the characteristic impedance for even and odd mode.

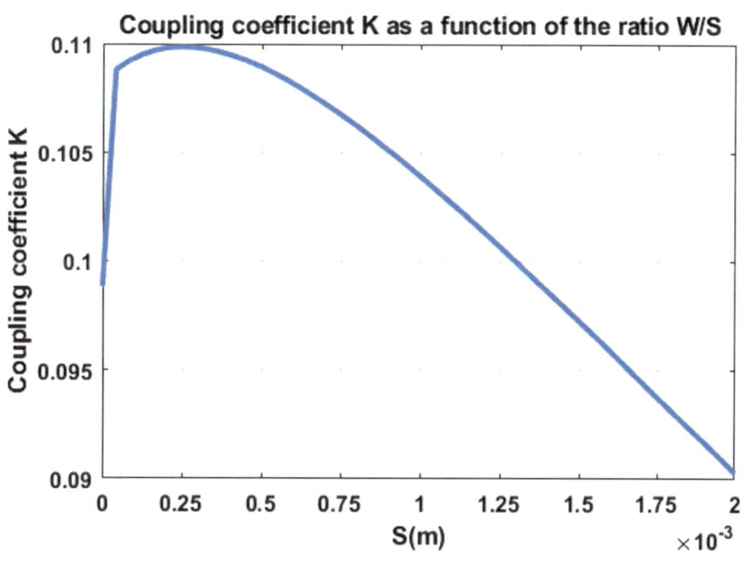

Figure 10.
Variations of the coupling coefficient.

$$Z_{co} = \frac{\omega}{\beta_o}\left(L_1 - \frac{L_m}{R_e}\right) = \frac{\beta_o}{\omega}\left(\frac{1}{C_1 - R_o C_m}\right) \tag{29}$$

For the coupling coefficient K, it is given by the following formula:

$$K = \frac{Z_{ce} - Z_{co}}{Z_{ce} + Z_{co}} \tag{30}$$

Figures 9 and **10** show the variations of Z_{co}, and Z_{ce} as a function of the ratio W / S, and the variation of K the coupling coefficient as a function of the gap S between the coupled lines respectively.

7. Conclusion

The analysis presented in this chapter has made it possible to modelized the most important characteristic parameters of a directional coupler in CPW (coplanar waveguide) technology, Such as the characteristic impedances for different modes of propagation: odd and even in quasi TEM mode by using the general theory of coupled lines. The directional coupler studied is based on two micro-coplanar lines in which we utilize the integral equation method (solved numerically by using the MOMs) associated with the Green's function, and conformal mapping method to describe the different characteristics of one and two micro-coplanar line.

Acknowledgements

I must start by thanking my professor Naima Amar Touhami for her encouragement. I also want to thank the author service manager Jasna Bozic for her assistance and all members of IntechOpen.

Author details

Anouar Acheghaf* and Naima Amar Touhami
Faculty of Sciences, Abdel Malek Esaadi University, Tetouan, Morocco

*Address all correspondence to: anoua.miamo@gmail.com

References

[1] K.C. GUPTA, RAMESH GARG AND INDER. BAHL AND MAURIZIO BOZZI. Microstrip lines and slot-lines, Artech house third edition Washington 2013. ISBN: 9781608075362 1608075362.

[2] R.W. JAKSON, Consideration in the use of coplanar waveguide for millimetric wave integrated circuits, IEEE TRANS. On MTT, vol. MTT-34, no 12, December 1986, pp 1450–1456. DOI: 10.1109/TMTT.1986.1133562.

[3] WALTON C. GIBSON, The method of moments in electromagnetics. Chapman & hall/CRC. ISBN-13: 978–1–4200-6145-1.

[4] A. BENGHALIA. Contribution to the two-dimensional modeling of micro-coplanar lines on a semiconductor substrate [thesis], INP TOULOUSE, September1989.

[5] W.T. WEEKS, Calculation of coefficients of capacitance of multi-conductor transmission lines in the presence of a dielectric interface. IEEE trans on MTT, volume MTT-18, No 1, Jan 1970 pp.35–34. ISBN: ISBN: 1–55899-375.

[6] J.R. MOSIC, Microstrip structures: analysis by means of integral equations [thesis], Federal Institute of Technology of Lausanne 1983.

[7] POSTOYALKO, Green's function in a region with inhomogeneous, isotropic dielectric media. IEEE vol, MTT-34, nov.1986, pp 1092–1095. DOI: 0018–9480(96)00469–3.

[8] MASANORI KOBAYACHI, Analysis of the micro-strip and the electro-optic light modulator, IEEE TRANS. On MTT, vol. MTT-26, No 2, February 1978, pp 119–126.

[9] MASANORI KOBAYACHI, Green's function technique for solving anisotropic electrostatic field problems, IEEE TRANS. On MTT, vol. MTT-26, No 7, July 1978, pp 510–512.

[10] ROBERT E. COLLIN, Field Theory of Guided Waves. Edition Mc Graw Hill, New York, 1962. ISBN: 978–0–879-42237-0.

Chapter 6

Improving Product Quality through Functional Analysis Approach: Case of Dual Axis Solar Tracker

Emmanuel Karabo Mpodi, Zeundjua Tjiparuro and Oduetse Matsebe

Abstract

Product quality determines how well a product meets the customer's requirements. One way of measuring and ensuring that the product's quality is achieved is through incorporating the functional analysis approach in the design process of the product, especially at early stage of lifecycle. A case study involving the design of a dual axis solar tracking system is used to illustrate the approach. In the study, the designed solar tracking concept was compared to existing mechanisms. The designed concept was found to be, generally, less complex than existing models.

Keywords: design, quality, complexity, solar energy, functional analysis

1. Introduction

Since the industrial age, product development has evolved greatly from the primitive craft approach in which design and manufacturing were interlinked to the new enhanced approach in which design and construction are separate. The product development process is often achieved in six (06) steps as illustrated in **Figure 1**. Whereby **Strategic definition** is the identification of the need or a niche in the market, **Research** is the competitive analysis of products related to the one being developed in order to understand the dynamics of the product market and demands, **Product design and manufacturing** transforms ideas into functional objects (physical or virtual). **Production evaluation and selection** is the stage at which the choice of the engineered product is done influenced by factors such as; cost, user preference, product quality, etc. **Use and recycling of product** is the stage at which the product is put to use, encounters tear and wear, it is repaired until it reaches the end of life after which it is disassembled and some parts might be used for other purposes. Finally, **Market feedback** is the input from the market to help improve future generations [1].

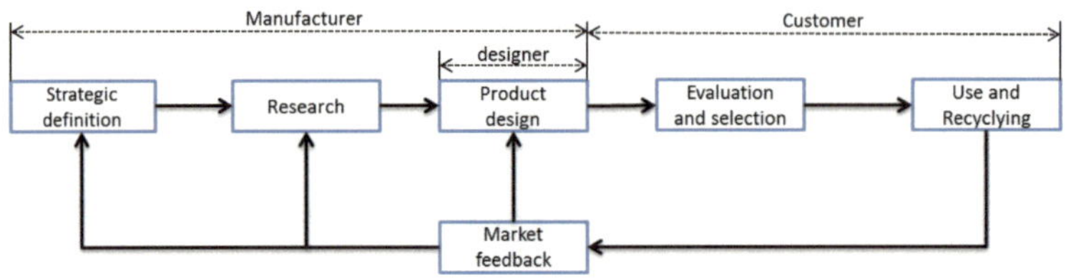

Figure 1.
Product lifecycle [1].

2. Design process model

There are many design models. For instance, French's descriptive model has four stages, namely; a) analysis of problem, b) conceptual design, c) embodiment of schemes, and d) detailing [2]. Cross' model is four staged too. The four stages are; a) Exploration, b) generation, c) evaluation and d) communication [3]. So is Ullman's model consisting of; a) product planning, b) conceptual design, c) product development and d) product support [4]. The stages of all the three models above are nearly similar and apply the general framework given in **Table 1**. Other models comprise of Axiomatic design, the VDI model, quality loss function, and quality-function deployment, and many others [5].

2.1 Design quality

Quality is defined as the ability of the supplier/producer to meet the specified and measurable requirements of the customers. From this definition of quality,

Design stage (descriptive)	Sub-stages (prescriptive)	Relative techniques
Planning	Deriving customer's needs	Questionnaires, usability lab studies, ethnographic field studies, etc.
	Setting design objectives	Checklists, objective and key results (OKR), Specific, measurable, actionable, realistic and time-based (SMART) framework, mind maps, etc.
Generation	Functional analysis	Function-means tree, Becoming-the-flow, Converter-Operator-Transmitter-Control model (COTC) Bond graph model, etc.
	Setting technical specification	Quality function deployment, design for assembly, design for manufacture, theory for inventive problem solving (TRIZ) matrix etc.
	Generating design alternatives	Morphological analysis, brainstorming, biomimetic, design by analogy, 6-3-5, etc.
Evaluation	Testing and validation of product	Simulations, mockup, prototyping, mathematical models, miniaturised models etc.
	Product improvement	Value engineering, Failure mode effect, Fault tree analysis, design for environment, design review, Strength, weakness, opportunities and threads (SWOT) analysis, Pros-cons analysis etc.
Documentation	Detailed design	Technical drawings, designs portfolios, procurement plans etc.
	Design documenting	Design database, patents, product manuals, design report etc.

Table 1.
Summary of theoretical design framework.

design quality is then defined as a practice of ensuring that products developed in a process of design meet the expectation of customer (without imposing any harm to the social and natural environment of society). It is important to control and monitor quality of products in order to minimise cost, resource, time and relative environmental impact of product development.

A five-level hierarchy of design quality was proposed by reference [6]. The attributes outlined in the reference are namely; functionality, reliability, usability, maintainability, and creativity. Functionality of products is considered paramount in controlling, managing, and ensuring that high quality designs are achieved [6].

Simplicity and complexity are also concepts used to define quality of design products. Simplicity is the exact converse of complexity. Simplicity of an artefact is defined as the use of the lowest possible number of lines, shapes, components, etc. without compromising its functional requirements.

2.1.1 Functional analysis

Functional analysis transforms customer's requirements into functional means (physical components). In the approach, the designer surveys the prospective customer's market to develop a product that is suitable for their need. Often, simpler and competitive products than existing ones are realised by this approach [7].

2.1.2 Design complexity

Design complexity is a field in design engineering which focuses on analysing and managing uncertainties of designs (i.e. process and product) due to many interwoven elements and attributes which make an object difficult to understand. Managing complexity in design is important as it reduces effort and resources used when developing products. Design complexity metrics measure a number of design aspects such as; structural complexity: (i.e. physical arrangement and interactions of constituting components), functional complexity: (i.e. number, variety, and interactions of basic and support functions), behavioural complexity: (i.e. predictability and understand-ability of product's behavioural in the field) [8].

Three complexity metrices exist. Bashir and Thompson (1999) developed a design complexity metric system that uses a functional analysis approach [9]. The devices are broken down from basic to advanced functions. This approach considers a linear relationship of functions at each level, but the number of assemblies and components in a device are neglected. Roy et al. (2010)'s complexity metric method was formulated to address the demand of the device with regard to the commonality of components used to construct the device [10]. Whereby product commonality is the number of parts being used for more than one product and is measured for all product family. Keating (2000) developed a complexity metric system which is based on the number of components and their interaction in a device [11]. **Table 2** gives a summary of the metrics and reasons for disregarding some and choosing one.

3. A case study: design and complexity evaluation of dual axis solar tracking concept

To illustrate how the functional analysis technique can be used to remove complexity and ensure that product quality is achieved at the early stages of product development of an engineered system, a design case study for the design of a dual axis solar tracking system is used. **Figure 2** gives the general design framework used.

Reference	Description	Formula	Comments
[9]	Number of functions	$C = \sum_{j=1}^{l} F_j k_j$ Where; C = complexity L = number levels F_j = number of functions at level j K_j = weight of level j; 1,2,	Not selected because only a few of the publications reviewed in this study disclosed the functional analysis of their designs.
[10]	Demand	$C = {}^{d_i}/_{d}$ Where; d_i = demand of part variant d = total demand of product	This approach is based on the availability of product components in the market. Therefore, this method is not relevant for use in this study.
[11]	Number of components and interaction	$C = M^2 + I^2$ Where; M = Module/components I = interactions	Since most publications describe their devices in terms of assemblies, components, and their interactions, this method was found to be the most suitable for this research.

Table 2.
Design complexity metric system.

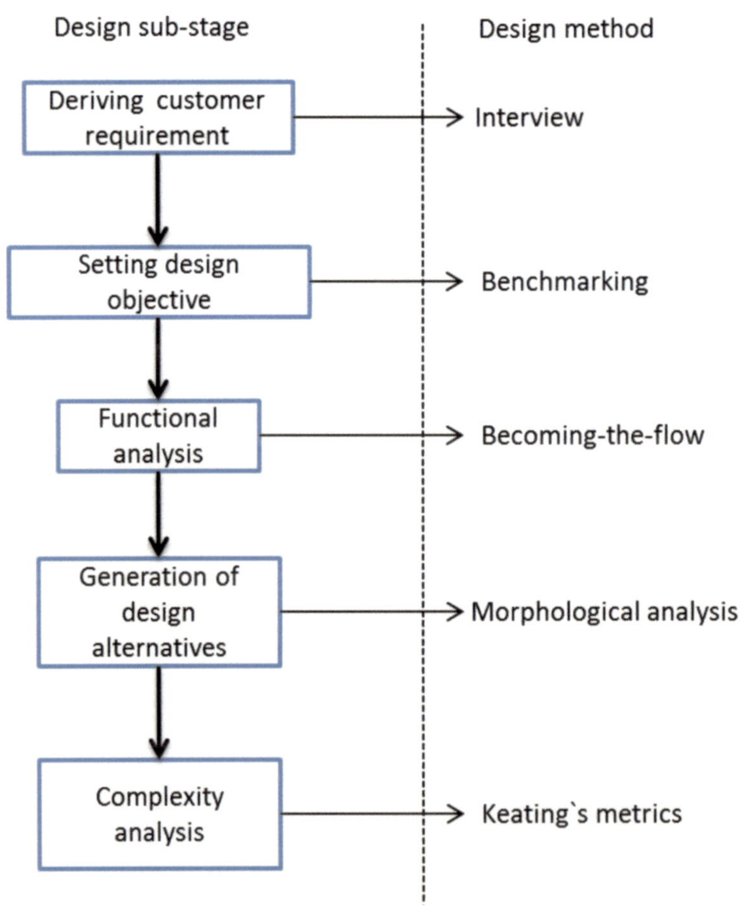

Figure 2.
Design approach for developing a solar tracker.

3.1 Deriving customer's requirements

The requirements of a dual axis tracker were established from an interview conducted with a facility technician at the Phakalane solar plant (Botswana). This was to understand the requirements of a dual axis solar tracker from an expert. The following questions divided into two categories, namely, functional and non-functional aspects were asked in the direct interview:

Functional aspects: The following six (06) questions were asked under functional aspects.

1. What defines the best performing solar tracking, in terms of its;
    a. Level of efficiency
    b. Power consumption
    c. Tracking accuracy
2. How is the solar tracking going to be operated i.e. manually, semi-automatic or automatic?
3. Under which environmental conditions is the device going to operate?
4. Is the device profitable (i.e. what is it payback)?
5. What level of maintenance and repairing is required?
6. What type of technology is required to operate the device?

Non-functional aspects: the following four (04) questions were asked under this non-functional aspect.

1. What method of waste disposal will be used after product life cycle?
2. Is the 3Rs (reuse, reduce and recycle) approach embedded in the product?
3. How will the operation of the device affect wildlife, birdlife and water sources?
4. What level of aesthetics is required for the system?

The requirements described in **Table 3** were identified during the interview.

3.2 Setting design objectives

Firstly, the Universal Track Racks™ by ZomeWorks (in **Figure 3**) was used to come up with the basic functions of a solar tracking system due to its popularity. The tracking system uses two or four *(if is dual axis)* identical cylinders on the edges of a panel frame. These contain a working thermo-fluid (normally a refrigerant). As the position of the sun changes with time, one cylinder receives more thermal power than the other. Due to this, the refrigerant expands and flows to another cylinder through a duct. From this process the function identified is the ability to detect the new position of the sun at a reference point. As the fluid accumulates in the other cylinder, there is difference in weight of the two cylinders. Since system is

Design requirements	Description
Low tracking error	A highly efficient tracking is the one that can position towards the sun with relatively high accuracy, for an improved energy output.
Low energy consumption	For an economically feasible product, the tracking device should consume as little energy as possible or use a mechanism which saves energy.
Fully automated	A machine with little human interface of daily operation, but with ease of use by an operator.
Operational in Array setup (On-Grid)	The tracker should be used on national electrical grid-connected PV system.
Optimum Power output	The solar tracking device should generate enough power either equal or slightly lower than the theoretical expectation, for economical and functional viability.
Optimum Payback period	For an economically viable system there is a need that it has a lower payback period as the profit will be realised in the early period lifetime of the machine.
Environmentally Friendly	Solar energy aids in reducing pollution emission. Therefore, the device should not harm its surroundings e.g. ecological system, water sources, wildlife and birds through generation of toxic waste materials
Aesthetically appealing	Growth in the use of renewable energy technology has led to an increasing interest in many people to comprehend the technology. Therefore, the solar tracking should be aesthetically attractive to attract tourists (i.e. technological tourism)

Table 3.
Identified design requirement for dual axis solar tracker.

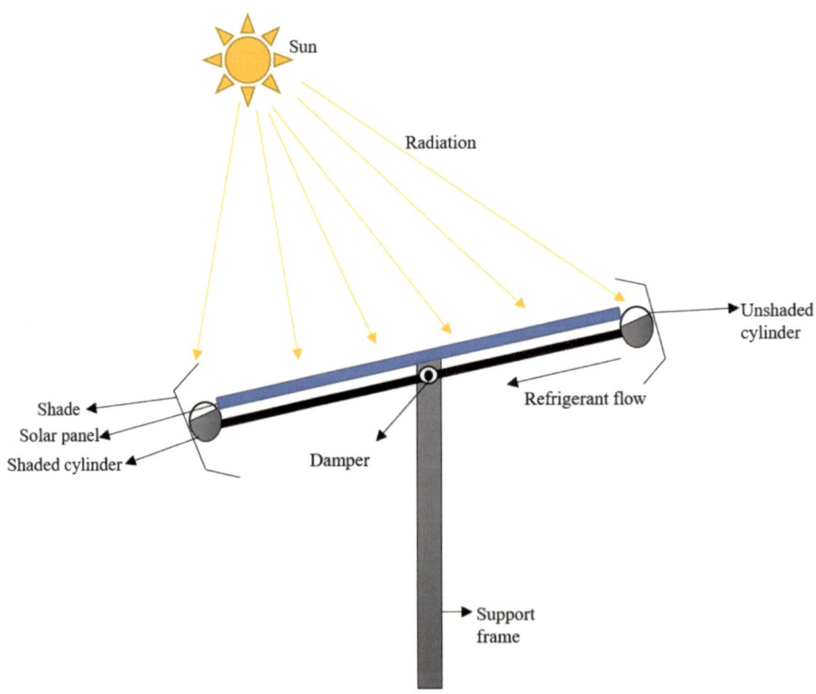

Figure 3.
ZomeWorks passive solar tracker [12].

designed to detect imbalance through a centre pivot, then motion is generated by this effect. To avoid shock on the system, a damper is used to guide and regulate; as other components rotate towards the desired position [12]. From this process four main sub-functions can be summarised as follows;

a. To precisely determine the position of the sun.

b. To calibrate the positioning mechanism.

c. To generate the tracking motion.

d. To monitor tracking effect.

3.3 Functional analysis

Form follows functions approach was used at this stage. That is, functions establishment should be independent of geometrical state to broaden the solution space [4]. A function is a task that transforms input to output in the system [13]. Therefore, functional modelling is a framework that relates functions in a flow, processes, and the operations to a system.

Retrieval questions	Description	Inputs identified	Type of input (energy, material or signal)	Output
Why is there a need to track the sun for PV application?	There is change in position the sun (triggers the need to measure the change in sun position).	Sun position	Signal	• Sun position • Signal for control
	To increase output of PV by tracking (PV generates electrical energy from sunlight directly)	PV system Solar energy	Material Energy	• Electrical energy • PV • Solar energy
Can PV system rotate on itself?	There is a need for support structure to provide facilitate motion and solid orientation (the support structure is coupled to the PV)	PV system coupled to the support structure	Material	• PV system + Support structure
How is automation of the system going to be achieved?	The level of interaction with human is low. That is the user only monitors the machine at time of maintenance and unforeseen operations	User	Signal	User
Is the system environmental conditions proof?	The environmental conditions such as wind, rain and cloud shade will affect the tracker	• Wind load • Rain load • Measure of wind • Measure of rain • Measure of cloud shade	• Energy (Wind and rain loads) • Signal (Measure of rain, wind and cloud shade)	• Wind load • Rain load • wind • rain • cloud shade
How is the energy going to be minimised?	This is based on choice of input energy and mechanism (there selection of mechanical energy for providing torque with recycling of waste energy and electrical energy used for power calibration devices can minimise energy)	• Mechanical Energy • Electrical Energy	Energy	Waste energy (heat and noise)

Table 4.
Thought aid process applied for becoming the flow.

Using *"to become the flow"* heuristic approach, the transparent box model was developed to identify the *"function chains"* (i.e. related tasks often performed by a single physical component). **Table 4** shows a thought aid process (e.g. the retrieval questions) used in the *"becoming the flow"* approach. *"Becoming the flow"* approach is based on the flow of energy, signal and material in a system. The first step engaged in this process was to identify the inputs, outputs and following their interactions in the solar tracking system. Inputs in this study are defined as fundamental *"causes"* that ensure that the overall function of the system is performed. Output is the *"effect"* produced in the system i.e. these include; desired and undesired effects. Some of the inputs remain unchanged, while others change (i.e. they are consumed) in a process carried out by a system. These inputs were traced until they exit the system as outputs. To identify inputs and outputs the following guideline were used, a consideration of the requirements (i.e. functional aspect of system), environmental conditions and designer's understanding of the problem.

Transparent box model of a solar tracking device is shown in **Figure 4**. In this model, energy, material, and signal are traced from input to their relative output state. The model was used to identify function chains to achieve relevant tasks. In the stated figure the (SS.) stands for support structure, (tor.) is torque, (sys.) is system, (Ener.) is energy, (mech.) is mechanical, (elec.) is electrical, (Pow.) is power, (Enviro.) is environmental, (Pos.) is position and (Prot.) is protection [14].

3.4 Generation of design alternatives

A morphological Chart was deployed to perform this transitional process, i.e. to present design alternatives generated in this research. Firstly, the function chains identified with the aid of transparent box model were listed in the column of morphological chart (grid). Then possible alternatives (i.e. these are physical components available in market) to perform the tasks of the function chains were identified. Through brainstorming, the grid of the morphological chart was filled by noting (with text) ideated alternatives alongside their relevant function chains (i.e. on the row of the function chain). For example, two alternatives; electronic

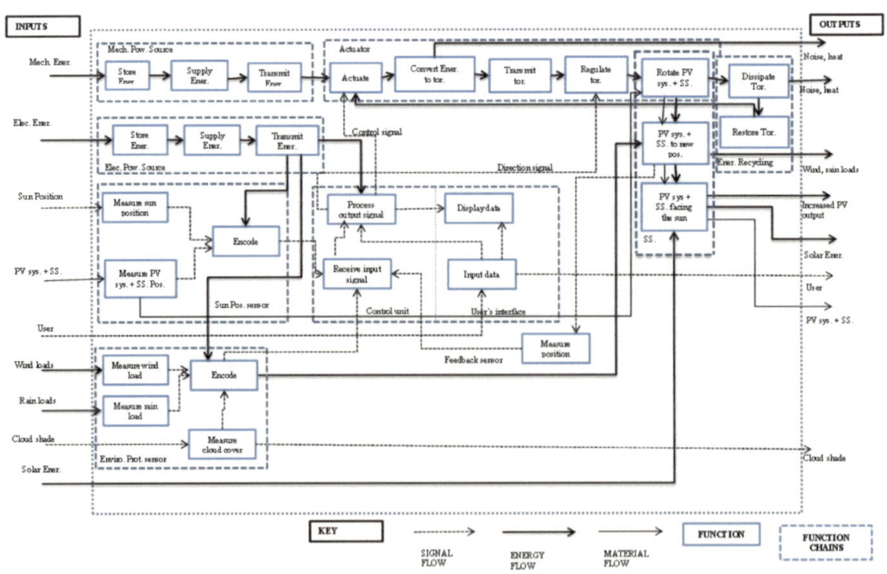

Figure 4.
Transparent box model of a dual axis solar tracking [14].

Function chain (Func.)	Alternatives (Alt.)				
	Alt 1	**Alt 2**	**Alt 3**	**Alt 4**	**Alt 5**
Sun position sensor (Sun Pos.)	Photo sensors	Real-time clock (RTC)	Camera	Global positioning system device (GPS)	RTC+ photo sensor (Hybrid)
Power source (elec.)	Mini PV panel	Grid electricity			
Power source (mech.)	Solar engines	Spring system	Gravity engines		
Control unit (CU.)	Micro controller	Personal computer (PC)	Programmable Logic controller (PLC)	Field Programmable Gate Array (FPGA)	
Actuator (Act.)	Hydraulic cylinder	Pneumatic cylinder	Motor and gearbox	Stepper motor	
User's interface (UI.)	Keypad and LCD screen	Safety switch and LED flashlight			
Support structure (SS.)	Cable mount	Parallel kinematics device (PKD)	Rotating platform (RP)	Polar mount	Counterbalance mount (CBM)
Energy recycling system (ER.)	Spring system	Piezoelectric system	Spring return fluid power actuators	Energy recovery wheel (ERW)	
Feedback sensor (FS.)	Inclinometer	Accelerometer	Magnetometer	Gyroscope	
Wind sensor (WS.)	Electronic Anemometry (EA)	Airflow sensors			
Rain sensor (RS.)	Weighing precipitation gauge (WPG)	Optical rain gauge	Water Sensors		
Cloud sensor (CS.)	Optical sensor	Ceilometer			

Table 5.
Morphological chart present design alternative [15].

anemometry (Alt 1) and airflow sensor (Alt 2) were brainstormed for the function chain; wind sensor (**Table 5**) [15].

$$\text{Possible combinations} = \prod_{i=1}^{n} O \tag{1}$$

$$= 5 \times 2 \times 3 \times 4 \times 4 \times 2 \times 5 \times 4 \times 4 \times 2 \times 3 \times 2$$

$$= 4,147,200$$

The evaluation measures formulated at the planning stage of the design process were then deployed to judge the alternatives. As a way of guiding selection of best alternatives, that will be used to develop a concept. Some of the evaluation measures, which are normally used for evaluation of concepts, are defined below:

- Serviceability/maintainability: This attribute describes the timeliness, relative cost and availability of skilled personnel in the local areas to carry out replacement and/or repair of components.

- Reliability: the ability to maintain an expected functional behaviour at all times and under specific conditions.

- Interfacing/compatibility: the ability of the component to be useable with different configurations and strategies to achieve the desired function.

- Scalability: can a component be easily down or up sized for a specified application.

- Cost: the price value of a single component will affect the total cost of device hence its economic feasibility.

- Availability: ease of access of a component locally or less difficulties in sourcing it.

Evaluation of alternatives was then carried after a five-point Likert scale was established. Then each alternative was scored against the evaluation measure in a relevant manner (i.e. according to the knowledge and discretion of the designer). Points scored by each alternative were aggregated, and the alternative scoring high points were ranked as first choice (refer to **Tables 6** and 7) [15].

Lastly, a concept was developed from aggregating the best-selected alternatives. This resulted in the final design which was modelled using a SolidWorks® platform (**Figure 5** shows the developed concept).

3.5 Complexity analysis

The analysis was carried out by comparing the existing systems' design complexity with the developed concept. In the comparison, the approach used in reference [11] was adopted. This approach uses modules and interactions between the modules to compare design products. A typical Keating's model is given in **Figure 6** whereby the number of components/modules (M), and number of interactions (I), in the design are counted and the inherent complexity computed using (Eq. (2)).

$$C = M^2 + I^2 \tag{2}$$

Table 8 shows a complexity metrics of systems developed in the period, 1997-2017. The average complexity of these systems was found to be 221.43 in this research.

Figure 7 shows a diagrammatic embodiment design of the designed system. Plotting the complexity index of the developed concept against the complexity values of the existing systems give **Figure 8**. The trend illustrated the graph shows a relatively constant increasing pattern at the beginning of the study period up to the year 2005-2007. Generally the system designed is more complex when compared with developed between 1997 and 2004. While from 2005 to 2011 the existing

Func.	Criteria	Alt 1	Alt 2	Alt 3	Alt 4	Alt 5
Sun Pos.		Photo sensor	RTC	Camera	GPS	Hybrid
	Serviceability	4	5	1	3	4
	Availability	5	4	1	3	4
	Interfacing	4	5	1	3	5
	Reliable in cloudy weather	3	2	4	1	4
	Point scored	**16**	**16**	**7**	**10**	**17**
	Rank	**2**	**2**	**4**	**3**	**1**
Electric.		Mini PV panel	Grid			
	Serviceability	5	4			
	Availability	5	5			
	Reliability	4	3			
	Points scored	**14**	**12**			
	Rank	**1**	**2**			
Mech.		Solar engines	spring system	gravity engines		
	Serviceability	5	5	4		
	Availability	4	5	2		
	Reliability	5	2	2		
	Points scored	**14**	**12**	**8**		
	Rank	**1**	**2**	**3**		
Act.		Hydraulic cylinder	Pneumatic cylinder	Motor and gearbox	stepper motor	
	High response	3	4	2	5	
	Controllability	4	4	2	4	
	Interfacing	4	5	2	3	
	Minimal energy consumption	3	4	3	5	
	Compatible to support structure	5	5	3	4	
	Points scored	**19**	**22**	**12**	**21**	
	Rank	**2**	**1**	**3**	**2**	
CU.		Microcontroller	PLC	FPGA	PC	
	Interfacing	5	3	2	3	
	Accuracy and Precision	3	5	4	4	
	Availability	5	4	3	5	
	Serviceability	5	4	2	2	
	Adaptability to control	4	5	5	5	
	Point scored	**22**	**21**	**16**	**19**	
	Rank	**1**	**2**	**4**	**3**	

Table 6.
Evaluation of design alternatives for solar tracking [15].

Func.	Criteria	Alt 1	Alt 2	Alt 3	Alt 4	Alt 5
UI.		Keypad and LCD screen	LED and switch			
	Accessibility of information	5	3			
	High alarm rate	4	4			
	Compatibility	5	4			
	Points scored	**14**	**11**			
	Rank	**1**	**2**			
SS.		Cable mount	Polar	Parallel mech.	CBM	RP
	Optimal land coverage	4	3	2	5	1
	Versatile utility	3	4	2	5	1
	Assemble-ability	5	3	1	5	3
	Optimal material consumption	5	3	1	4	2
	Robust mechanical	1	4	5	3	5
	Points scored	18	17	11	22	12
	Rank	2	3	4	1	5
ER.		Springs	Piezoelectric	Spring return fluid power actuators	ERW	
	Compatibility to control	5	3	5	1	
	Ease of use	4	1	5	2	
	Maintainability	5	2	4	1	
	Availability	4	2	5	1	
	Points scored	**18**	**8**	**19**	**5**	
	Rank	**2**	**3**	**1**	**4**	
FS.		Accelerometer	Inclinometer	Magnetometer	Gyroscope	
	Interfacing	4	5	1	2	
	Cost	5	5	1	1	
	Availability	5	5	2	2	
	Points scored	**14**	**15**	**4**	**5**	
	Rank	**2**	**1**	**4**	**3**	
WS.		Electronic anemometry	Airflow sensor			
	Interfacing	4	4			
	Availability	3	5			
	Scalability	3	4			
	Cost	4	5			
	Points scored	**14**	**18**			
	Rank	**2**	**1**			

Func.	Criteria	Alt 1	Alt 2	Alt 3	Alt 4	Alt 5
RS.		Weighing gauge	optical gauge	water sensor		
	Interfacing	1	1	5		
	Availability	2	3	5		
	Scalability	2	4	5		
	Cost	2	1	5		
	Point scored	7	9	20		
	Rank	3	2	1		
CS.		Optical sensor	Ceilometer			
	Interfacing	5	2			
	Scalability	4	3			
	Availability	5	1			
	Cost	5	1			
	Points scored	19	7			
	Rank	1	2			

Table 7.
Continuation of evaluation of design alternatives [15].

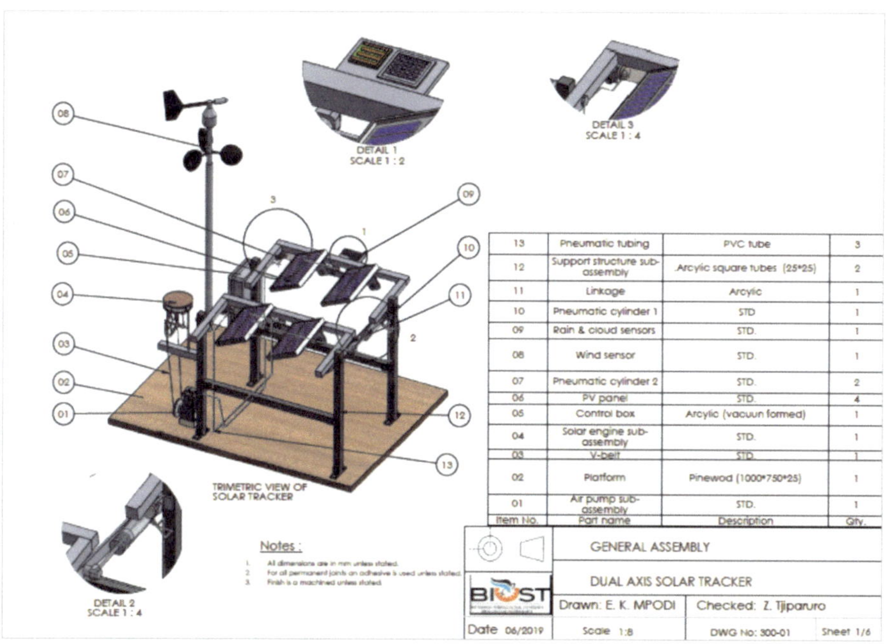

Figure 5.
General assembly drawing of the solar tracking concept developed.

systems are more complex the concept developed in this research study. For period between 2012 and 2017 the system developed and existing system are generally equal in complexity. The pattern was realised because of the advancement which were made to the dual axis tracking such as including weather intelligent features

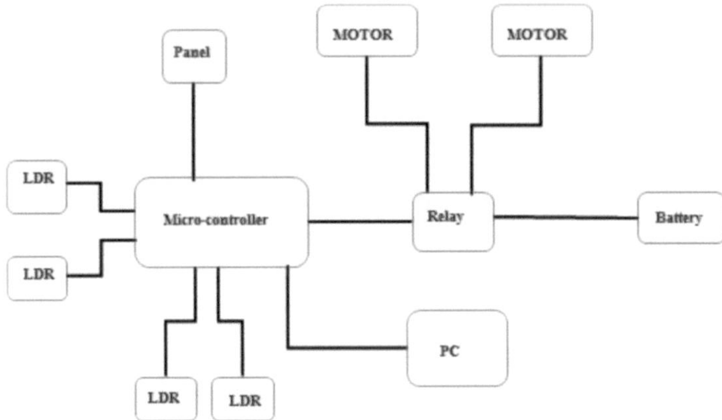

Figure 6.
A block diagram showing module and interaction of system developed by Akbar et al. (2017).

Reference	M	I	C
[16]	4	6	52
[17]	7	8	113
[18]	5	4	41
[19]	10	11	221
[20]	15	18	549
[21]	4	3	25
[22]	6	8	100
[23]	10	10	200
[24]	14	16	452
[25]	13	11	290
[26]	10	10	200
[27]	10	9	181
[28]	10	12	244
[29]	12	11	265
[30]	9	10	181
[31]	10	11	221
[32]	11	13	290
[33]	14	15	421
[34]	9	10	181
[35]	9	11	202
[36]	11	10	221
Average			221.43

Table 8.
Design complexity study of existing solar trackers.

(wind and rain shield systems). In summary the developed system, firstly, falls within the average complexity of existing systems, and secondly, it is 10% less complex than the existing systems.

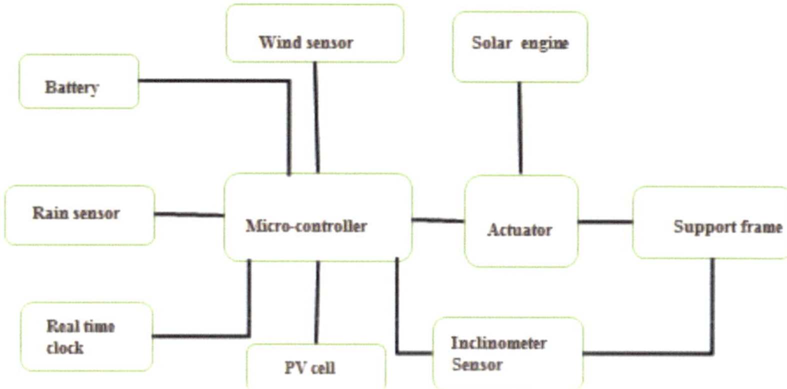

Figure 7.
Embodiment diagram of a solar tracking concept developed.

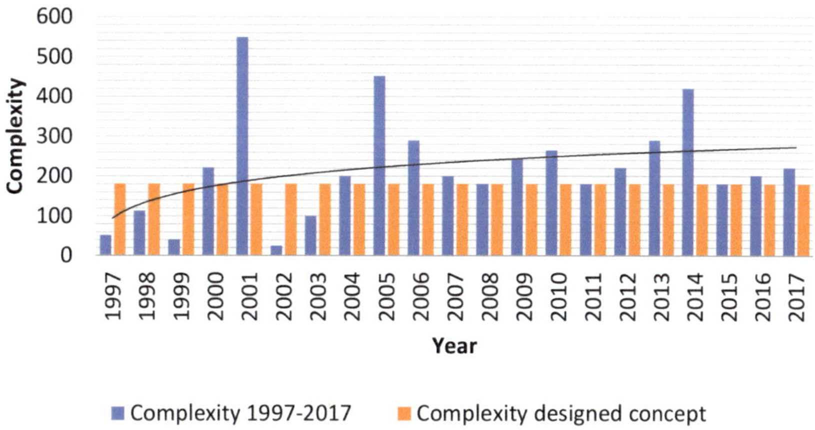

Figure 8.
A comparison of design complexity for the developed concept and existing mechanisms.

4. Conclusion

In this chapter a design of a dual axis solar tracker was used to describe a way of enhancing product's quality, during the early stage of product design. A design and complexity analysis undertaken resulted in a less complex solar tracker. The developed concept was evaluated against the existing solar tracking systems. Therefore, carrying out an analysis of complexity on system at an early stage of product design is important in improving the product functionality and simplicity factor. Consequently, this will relatively reduce the product's cost and design effort.

Acknowledgements

I would like to thank Botswana International University of Science and Technology for technical and financial assistant.

Conflict of interest

The author(s) declared no potential conflict of interest in regard to this research, authorship and/or its publication.

Author details

Emmanuel Karabo Mpodi*, Zeundjua Tjiparuro and Oduetse Matsebe
Department of Mechanical, Energy and Industrial Engineering, Faculty of Engineering and Technology, Botswana International University of science and Technology, Palapye, Botswana

*Address all correspondence to: emmanuel.mpodi@studentmail.biust.ac.bw

References

[1] Ulrich K. Product Design and Development, McGraw-Hill. 2008.

[2] French M J. Conceptual Design for Engineers. 3rd ed. London: Springer Berlin Heidelberg; 1985.

[3] Cross N. Engineering Design Methods: Strategies for Design Products. 3rd ed. Chichester, England: John Wiley; 2008.

[4] Ullman D. Mechanical Design Process. 4th ed. McGraw; 1992.

[5] Amersfoort J. VDI design guide: Comprehensive Guide to Help You Design VMware Horizon, Based on Modern Standards. 1st ed. CreateSpace Independent Publishing Platform,; 2018.

[6] Bradley S. Designing for a hierarchy of needs. http://www.smashingmaga zine.com/2010/04/designing-for-ahie rachy-of-needs (2010, accessed 2 July 2020).

[7] Ahmed, A., & Ahmed, B. M. Product design and development by functional analysis. International Journal of Industrial Engineering and Technology. 2016; 8(1), 25-33.

[8] Ameri, F., Summers, J. D., Mocko, G. M., & Porter, M. Engineering design complexity: An investigation of methods and measures. Research in Engineering Design. 2008. Springer. London

[9] Bashir H, Thomson V. Estimating Design Complexity. Journal of Engineering Design. 1999;10(3):247-257.

[10] Roy, R., Evans, R., Low, M. J., & William, D. K. Addressing the impact of high levels of product variety complexity in design and manufacture. Journal of Engineering Manufacture. 2010;225: 1939-1950.

[11] Keating, M. Measuring design quality by measuring design complexity. Proceedings IEEE 2000 First International Symposium on Quality Electronic Design (Cat. No. PR00525) pp. 103-108. San Jose, CA: IEEE. Retrieved from 09/03/19/ ISQED.2000.838861

[12] ZomeWorks Corporation. Northern Arizona wind and solar. (2017). Retrieved 2018, from http://www.solar-eletric.com/lib/wind-sun/zomework-trackrack-brochure.pdf

[13] Tor, S., Britton & G, Zhang, W. Functional modelling in conceptual die design (n.d).

[14] Mpodi, E. K., Tjiparuro, Z., & Matsebe O. Review of dual axis solar tracking and development of its functional model, in T.-C. Jen, E. Akinlabi, P. Olubambi, & C. Augbavboa, (eds.) Procedia Manufacturing. 2019; 35, pp. 580-588. doi:10.1016/j.promfg.2019.05.082.

[15] Mpodi, E. K., Tjiparuro, Z, & Matsebe, O. Development of dual axis solar tracking concept using morphological analysis, in Rodrigo S. Jamisola, Jr. (ed.); BIUST Research & Innovation Symposium 2019 (RDAIS 2019); Volume 1(1), Palapye, Botswana, pp. 46-53. ISSN: 2521-2292

[16] Hoffmann, R., O'gallagher, J., Winston, R., O'gailaghert, J., & Winstont, R. High concentration low wattage solar arrays and their applications. 1997; pp. 394, 739.

[17] Chiang, L., & Jacob, J. Low cost sun-tracking photovoltaic panel. 1998

[18] Yousef, H. Design and implementation of a fuzzy logic computer-controlled sun tracking system. 1999.

[19] Helwa, H., &. Bahgat, A. Maximum Collectable Solar Energy by Different Solar Tracking Systems. Energy Sources. 2000.

[20] Fanney, A. H., & Dougherty, B. P. A. hunter fanney building integrated photovoltaic test facility. 2001.

[21] Yousif, C. Comparison study between the performance of tracking and stationary solar photovoltaic systems in Malta. 2002.

[22] Vilela, O. C., Fraidenraich, N., & Tiba, C. Photovoltaic pumping systems driven by tracking collectors. Experiments and simulation. Solar Energy. 2003; 74, pp. 45-52.

[23] Abdallah, S., & Nijmeh, S. Two axes sun tracking system with PLC control. Energy Conversion and Management. 2004.

[24] Piao, Z., Park, J., Kim, J., Cho, G., & Baek, H. A study on the tracking photovoltaic system by program type. 2005 International Conference on Electrical Machines and Systems. Nanjing, China.

[25] Mamlook, R., Nijmeh, S., & Abdallah, S. A programmable logic controller to control two axis sun tracking system. Information Technology Journal. 2006.

[26] Rubio, F., Ortega, M., Gordillo, F., & López-Martínez, M. Application of new control strategy for sun tracking. Energy Conversion and Management. 2007.

[27] Chun-Sheng, W., Yi-Bo, W., Si-Yang, L., Yan-Chang, P., & Hong-Hou, X. Study on automatic sun-tracking technology in PV generation. 2008.

[28] Sungur, C. (2009). Multi-axes sun-tracking system with PLC control for photovoltaic panels in Turkey. Renewable Energy. 2009.

[29] Barsoum, N., & Vasant, P. Transaction in controllers and energy simplified solar tracking prototype. 2010.

[30] Kassem, A., & Hamad, M. A microcontroller-based multi-function solar tracking system. 2011 IEEE International Systems Conference, SysCon 2011 - Proceedings.

[31] Eke, R., & Senturk, A. Performance comparison of a double-axis sun tracking versus fixed PV system. Solar Energy. 2012.

[32] Anusha, K., & Chandra, M. R. S. (2013). Design and development of real time clock based efficient solar tracking system. International Journal of Engineering Research and Applications (IJERA). 2013; 3(1), pp. 1219-1223.

[33] Kuo, K., Wang, J., Su, Y., Chou, J., & Jiang, J. A novel power benefit prediction approach for two-axis sun-tracking type photovoltaic systems based on semiconductor theory. Progress in Photovoltaics: Research and Applications. 2014.

[34] Ceyda A T and Cenk Y. Comparison of Solar Trackers and Application of a Sensor less Dual Axis Solar Tracker, Journal of Energy and Power Engineering. 2015. 10.17265/1934-8975/2015.06.006.

[35] Ray, S., & Tripathi, A. Design and development of tilted single axis and azimuth-altitude dual axis solar tracking systems. 1st IEEE International Conference on Power Electronics, Intelligent Control and Energy Systems, ICPEICES 2016

[36] Akbar, H., ISiddiq, A., & Aziz, M. (2017). Microcontroller based dual axis sun tracking system for maximum solar energy generation. American Journal of Energy Research, 5(1), (pp. 23-27).

Chapter 7

Taguchi Method as a Robust Design Tool

Coşkun Hamzaçebi

Abstract

Taguchi Method is a powerful technique to optimize performance of the products or process. Taguchi's main purpose is to reduce the variability around the target value of product properties via a systematic application of statistical experimental design which called robust design. Robust Design is an important technique for product manufacturability and product life. Taguchi simplified the usage of orthogonal arrays to setup experimental design. Thanks to this development, researchers and engineers saved both time and money. Furthermore, Taguchi proposed the usage of S/N ratio in order to measure the effects of factors on the performance characteristics. In this study a brief knowledge about the Taguchi Method is given. Orthogonal Arrays and S/N ratios are described. Summary of a case study is given.

Keywords: Taguchi method, robust design, orthogonal Array, S/N ratio, Pareto ANOVA

1. Introduction

The quality of a product is the result of production process. The desired properties of the product should be revealed at the design stage. Towards the end of the 1950s, Dr. Genichi Taguchi put forward many concepts and methods to improve quality which based on robust design.

Robust Design (RD) means the design of a product that causes no problem under any case. RD signifies designing of a product which can work properly under different circumstances [1].

One of the important developments of the manufacturing industry is related to the application of modern off-line quality control techniques in product or process engineering. Many of these quality techniques were shaped by W. E. Deming. Taguchi built his philosophy on them. Deming's main success has been to convince businesses that the production process should be controlled statistically in quality improvement. Taguchi went a little further back and said that quality will be achieved at the design stage before production. Taguchi's main purpose is to reduce the variability around the target value of product properties. To achieve this, the controllable factors that cause this variability must be identified and the product and production process must be designed according to these factors. Taguchi's strategy is a systematic application of Experimental Design (DOE) and analysis in order to improve or design product and process quality. This strategy includes experimental minimization of an expected loss function to determine the best product design (or process design) [2].

 IntechOpen

Taguchi observed that the most important reason for a product to be rejected is variability in product specifications. Improving quality is through reducing variability. Efforts for quality should be made for zero deviation and zero distortion. All quality experts, especially Shewart and Deming, have addressed the issue of variability. Taguchi in one of his articles [3] -by using the **Figure 1** which has given under the title "Who is the Better Marksman?"- indicated that it is a difficult problem to eliminate variability in the production process. In this example, both gunners fire ten shots. If the average position of Gunner's A is calculated, it will be seen that the average is very close to the target. On the other hand, marksman B's average is far from the target. However, his shots are very consistent. When the variability is calculated for both marksmen, it will be seen that the variability of the gunner B is much less. Those who are interested in shooting can easily say that while it is possible to correct B's shots with a small adjustment, it will take a lot of effort to make A a good shooter. Taguchi argues that production processes are also similar to shooters in this respect. While it is possible to easily adjust the B sniper-like processes, improving the A sniper-like processes will take a lot of time, maybe even huge investments.

Taguchi proposes a two-step process to reduce product variability. These steps are as follows.

- To produce the product with the best methods, technology and techniques
- To produce all products in the same way

In order to fulfill above issues, Taguchi divides the activities into two parts as On-Line Quality Control and Off-Line Quality Control. While on-line quality control covers the quality activities during and after the manufacture of the product, off-line quality control includes market research and quality activities carried out during the development of the product and production process. These activities are design studies carried out before production begins. Taguchi defines three stages such as system design, parameter design, and tolerance design both for product and process improvement.

The most important stage of product or process design in terms of quality improvement is the parameter design stage. At this stage, DOE method is used to

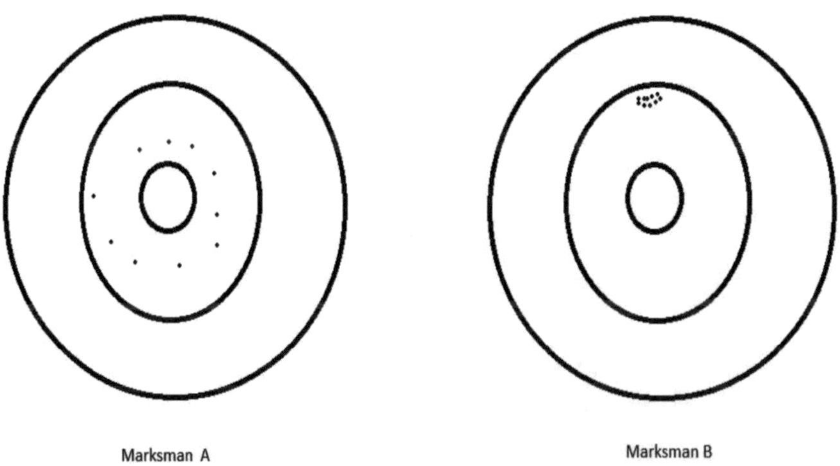

Figure 1.
Who is better gunner? (Adapted from Ref. [3]).

determine the factors affecting product performance and their effects on performance. The aim is to minimize the effect of effective factors on the product [4].

2. Literature review

RD is an important technique for product manufacturability and product life. Although the method was known by 1960's in Japan it has been used in USA by 1980's. Since its use in the USA industry in the 1980s, it has attracted a great attention from designers, manufacturers, statisticians and quality experts. Due to this success of robust design, a lot of researches such as master and PhD theses, scientific articles and case studies have been done to understand the method. Literature of Taguchi Method (TM) and RD is very large and it is still growing. When the literature is examined, it will be seen that Tagcuhi method is frequently used for the optimization of critical parameters of product and process in manufacturing industry and it gives useful results. **Table 1** presents some examples from last ten years publications about the manufacturing industry. It is important to note that TM has been applied to the service industry too. Antony [5] reports the potential applications of DOE in the service environment as follows.

- Identifying the key variables which influence the performance
- Identifying the service design parameters
- Minimizing the time to respond to customer complaints
- Minimizing errors on service orders
- Reducing the service delivery time to customers
- Providing a better understanding of cause–effect relationships between what we do and what we want to achieve
- Reducing cost of quality due to rework and misinformation that lead to bad decision-making

Recently publishings deal with the integration of TM and other approaches such as multicriteria decision making (MCDM), principal component analysis, numerical simulation, artificial neural network, and genetic algorithm. Sharma et al. [6] used the TM and PROMETHEE (which widely used MCDM tool) technique to obtain an optimal setting of process parameters for single and multi-optimization resulting in an optimal value of the material removal rate and tool wear rate. Kumar and Mondal [7] compared the results of experimental data on the electric discharge machining of AISI M2 steel by different optimization techniques such as TM, TOPSIS and gray relational analysis (GRA). Viswanathan et al. [8] aimed to investigate the effective factors in turning of magnesium alloy with physical vapor deposition coated carbide insert in dry conditions. To identify the optimal parameters setting, a combination of principal component analysis (PCA) and GRA has been conducted. Liu et al. [9] and, Land and Yeh [10] used both TM and ANSYS which widely used numerical simulation software in order to optimize and design injection molded products. Asafa et al. [11] presented integration of TM and artificial neural network (ANN) technique for the prediction of intrinsic stresses induced during plasma enhanced chemical vapor deposition of hydrogenated amorphous silicon thin films. Parinam

Article	Subject
Sekulic et al. [13]	Taguchi optimization methodology is applied to optimize cutting parameters in high-pressure jet assisted turning when machining Inconel 718.
Fei et al. [14]	The practical use of TM in the optimization of processing parameters for injection molding was reviewed. Also, integration of TM with various approaches including numerical simulation, GRA, PCA, ANN, and genetic algorithm (GA) were discussed.
Dave and Bhogayata [15]	The mix design of geopolymer concrete based on the target strength criteria by optimizing the proportions of the constituents using TM is presented.
Terzioğlu [16]	The factors which were effective in Thermoelectric Generators (TEG) used in the production of electrical energy a research is carried out by using TM to determine the performance effects.
Zhou et al. [17]	The effects of eight parameters on the value of borehole thermal resistance and internal thermal resistance are investigated. TM is carried out to obtain the optimal scenarios of parameters combination.
Hong [18]	A clustering approach based on TM for effective market segmentation is proposed. To select appropriate initial seeds, the use of TM as a tool is suggested.
Kumar et al. [19]	The objective of the article is to optimize and design nano-biosystem of Isradipine via novel bioenhancer (Rutin) loaded solid-lipid nanobioparticles using Taguchi design methodology.
Tiryaki et al. [20]	Taguchi design method for obtaining lower surface roughness values in terms of process parameters in wood machining is presented. Orthogonal arrays of Taguchi and the signal-to-noise (S/N) ratio is employed to find the optimal levels and to analyze the effect of process parameters on surface roughness.
Hamzaçebi [21]	TM was applied to determine the effects of production factors such as adhesive ratio, press pressure, and pressing time on the thermal conductivity of oriented strand board.
Alafaghani and Qattawi [22]	Taguchi's DOE is used to investigate the main effects of four processing parameters in the Fused Deposition Modeling (FDM) process; those are the infill percentage, infill pattern, layer thickness, and extrusion temperature.
Mitra et al. [23]	TM of robust optimization has been adapted along with DOE methodology and ANOVA to reduce the variability in the Ride comfort of a vehicle with respect to sprung mass of vehicle.
Çakıroğlu and Acır [24]	The optimization of the cutting parameters on drill bit temperature in drilling was evaluated by TM. TM was used to determining the settings of cutting parameters.

Table 1.
Some articles from the literature of the last ten years.

et al. [12] described integration of TM and Genetic Algorithms to optimize high transmission optical filter.

3. Robust design

Phadke defines the RD as an engineering methodology for improving productivity during research and development. Hence high-quality products can be produced quickly and at low cost [25]. The emphasis of RD is variability in product and process performance. Reducing variability will result in increased quality. The source of variability can be divided into two groups [26].

- Controllable factors: Factors determined by the manufacturer that cannot be changed directly by the customer,

- Uncontrollable factors (Noise factors): Factors that the producer cannot directly control and that vary according to customer use and environmental conditions.

Uncontrollable factors can be divided into three categories.

- External noise factors: factors such as environmental conditions, eg; environmental temperature, workers, different raw material piles etc.

- Intrinsic noise factors: time-varying factors, eg; deterioration, aging, discoloration, etc.

- Product-related factors: the difference in each product

Hence, RD means a design that has minimum sensitivity to variabilty of uncontrollable factors. Taguchi says that it is necessary to minimize the variability in the product or process by choosing the values of the controllable factors (parameters) optimally against the factors that create variability. The word robust in the statement of RD refers to uncontrollable factors which insensitive to environmental conditions such as moisture, dust, heat, different applications in customer use and differences in materials [27, 28]. The key to Taguchi Robust Design; instead of trying to control factors that cannot be controlled or that are too expensive to control, it is to determine the best values of controllable factors that will minimize their effects on the product or process [27]. RD provides answers to the following questions [29].

- How to reduce variability when the product is in customer use? How does a product consistently perform at the desired property and thus maximize customer satisfaction?

- How is the production process optimized?

As will be known, there are many factors that need to be determined and optimally adjusted in product and process parameter design stages. Moreover, many of these factors interact with each other. The most effective method to determine the effects of these controllable and uncontrollable factors on product and product performance is statistical experiment design. Through experimental design, it is possible to economically determine the effect of many factors on the product and to take precautions against factors that cause variability at the design stage. Therefore, we can say that the most important quality assurance method in Taguchi's off-line quality control system is DOE [30].

RD covers the parameter design and tolerance design steps of TM. System design consists of traditional research and development activities [31].

In order to realize RD, it is necessary to follow a systematic path. Implementation of the below steps are beneficial [26, 32, 33].

1. Determining the problem and organizing the experiment team

2. Determination of performance characteristics and measurement system

3. Determining the variables affecting performance characteristics

4. Establishing the monitoring design

5. Identifying controllable and uncontrollable variables and their levels

6. Identification of possible interactions

7. Selection of suitable orthogonal array and assignment of variables to relevant columns

8. Determination of loss function and performance statistics

9. Establishing the experiment and recording the results

10. Analysis of data and selection of optimum value of controllable variables

3.1 Determining the problem and organizing the experiment team

When a new product is to be developed, there is no need for any examination for the work to be done. If an existing product is to be developed, "why was this product chosen?" The question must be answered. Generally as an answer to this question; scrap, rework, warranty and service costs can be given. After the problem is determined, the team that will do the task should be formed. The team generally; It consists of experts of the problem of interest, DOE experts, senior management representative and people who will conduct the experiment. The other steps we try to explain below are carried out by this team.

3.2 Determination of performance characteristics and measurement system

The product may have one or more performance characteristics, so the selection of performance characteristics is important. The important point here is that the customer's view should not go unnoticed. Performance characteristics are the basis of the study. Determining the measuring system is the second step in this phase. Each of the performance characteristics may require different measuring systems.

3.3 Determination of variables affecting performance characteristics

Independent variables that affect the product performance characteristics should be determined. Previous experience and expertise are very important in this determination. Brainstorming, cause-effect diagrams and flowcharts are important tools to be used. Easily controllable independent variables are put in the group of control variables (CV) and the others into the group of uncontrollable variables (UCV).

3.4 Establishing the screening design

If the number of CV is large, it may not be possible to carry out the experiment in terms of time and cost. In such a case, there may be some variables that are believed to have no effect at the outset. Of course, making such a choice is difficult. Even after some variables are discarded, there is still the question of whether other variables are important. Screening design allows to get more realistic results with predetermined variables. In the sifter design, the level number is kept as low as possible, usually taken as two. The outputs are analyzed and junk CV is discarded. Significant CV is included in the main experimental group.

3.5 Determining the number and levels of CV and UCV

The number of levels of variables is determined by their characteristics. Thus, possible alternatives are obtained. Taguchi recommends selecting three or more test groups for each CV. Three or more test levels allow a nonlinear effect of CV on the performance characteristic to be revealed. Test levels should be chosen over a wide range so that the CV sequence covers a large region of the CV space. The next step is to determine the set of UCV. This cluster includes the values of the UCV that affect the performance variability the most or the product performance is insensitive. Due to physical impossibilities or lack of information, not all UCV can be included in the experiment. Therefore, it is important to represent all possible combinations of UCV in the experiment [34].

3.6 Identifying possible interactions

The definition of interaction can be as follows: If the effect of a factor on the response variable depends on the value of the other factor, it is said that there is an interaction between two factors as seen in **Figure 2** [30]. The interactions can have a significant impact on performance characteristics. Taguchi thinks that interaction is not that important. The reason of this; the view is that in order to detect the interaction, the experimenter has to control the two main effects, and the interaction does not contribute anything when one or more of the main factors are under control [33]. Taguchi and Wu [35] suggest that one of the following techniques should be applied to reduce the interaction effects.

1. Determining the performance characteristics by weight,

2. Determining the relationship between CV and its levels and making an adjustment accordingly,

3. Conducting an analysis for classified data, such as cumulative analysis.

However, the experimenter must have the necessary attention and knowledge. It is difficult to add all interaction factors to the experiment due to the high cost and time required. On the other hand, including interaction factors believed to be important in the experiment will increase success. The existence of interaction between two factors can be determined by graphical procedure.

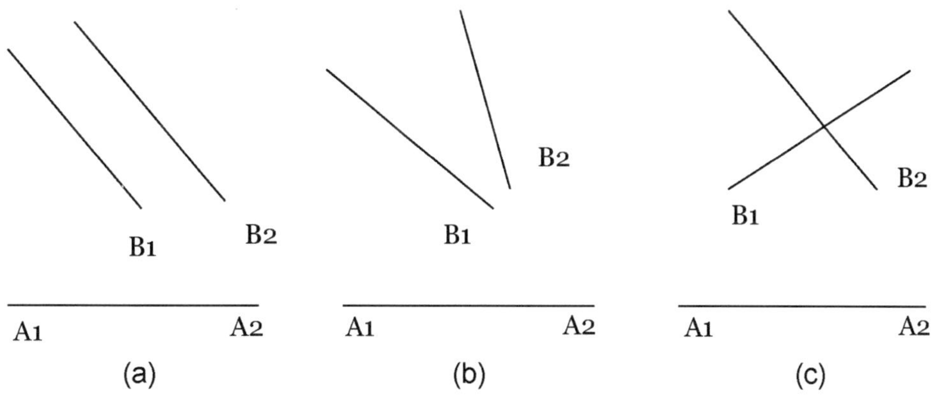

Figure 2.
Graphical representation of interaction between two factors. (a) No interaction, (b) Weak interaction, (c) Strong interaction.

3.7 Choosing appropriate orthogonal arrays

Orthogonal Arrays (OA) take us all the way to Euler's Greco-Latin squares. But in Euler's time they were not known as OA. At that time they were known as mathematical games, like 36 office workers' problems. OA is a matrix of numbers arranged in rows and columns. Orthogonal arrays have a balanced property which entails that every factor setting occurs the same number of times for every setting of all other factors considered in the experiment. In an OA, each row represents the levels of the selected factors in a given experiment, and each column represents a specific factor whose effects on the process performance or product quality characteristic can be studied.

The idea of using OA in DOE independently of each other is originated in the USA and Japan after World War II [36]. The first use of OA was in the 1930s by Fisher in England. Taguchi added three OAs in 1956. And in the following years, three OAs were added by the American NIST [31]. Taguchi makes use of OA in performing multivariate experiments with a small number of trials. Using OA significantly reduces the size of the experiment to be studied [37]. The use of OA is not exclusive to Taguchi. However, Taguchi simplified their usage. Taguchi developed tabulated standard OA and corresponding linear graphs. A typical OA table is shown in **Table 2**.

In this array the columns are bilateral orthogonal. In each column there are all combinations of factor levels with an equal number. There are 4 factors (A, B, C, D) and three levels of each. This design is called the L9 design. The letter L indicates the orthogonal array, and 9 the row number, in other words the number of trials [4].

One point we should pay attention to that how much the OA reduces the number of trials. Due to the full factorial design (2^k or 3^k), OA significantly reduces the number of attempts to be made in large numbers. For our example, $3^4 = 81$ trials are required, but only 9 trials will be done to achieve the same results. It is obvious that it will provide more convenience in larger series. **Table 3** highlights the convenience that OA provides in terms of the number of trials [37].

OA allows working economically and simultaneously with many variables that are effective in product mean and variance. Two different OAs can be selected for CV and UCV. Using statistical DOE techniques, suitable subsets for CV and CIA can be demonstrated. Taguchi suggests using OA in planning DOE optimization. The multiplicity of CV and the emergence of interaction require very careful attention in the selection of OA and assignment of CV to columns. Target in establishing CV

	A	B	C	D
1	1	1	1	1
2	1	2	2	2
3	1	3	3	3
4	2	1	2	3
5	2	2	3	1
6	2	3	1	2
7	3	1	3	2
8	3	2	1	3
9	3	3	2	1

Table 2.
L9 orthogonal Array.

OA	# of Factors and levels	Full factorial design trial number
L4	3 factors 2 levels	8
L8	7 factors 2 levels	128
L9	4 factors 3 levels	81
L16	15 factors 2 levels	32,768
L27	13 factors 3 levels	1,594,323
L64	21 factors 4 levels	4.4*1012

Table 3.
Frequently used OAs and full factorial design comparison.

OA	# of Row	# of Maximum factor	# of Maximum column			
			2 Levels	3 Levels	4 Levels	5 Levels
L4	4	3	3	—	—	—
L8	8	7	7	—	—	—
L9	9	4	—	4	—	—
L12	12	11	11	—	—	—
L16	16	15	15	—	—	—
L16'	16	5	—	—	5	—
L18	18	8	1	7	—	—
L25	25	6	—	—	—	6
L27	27	13	—	13	—	—
L32	32	31	31	—	—	—
L32'	32	10	1	—	9	—
L36	36	23	11	12	—	—
L36'	36	16	3	13	—	—
....	....	...	....	....	.....	....

Table 4.
OA information table.

matrix; It should be to setup a design where the most information can be obtained with the least effort. **Table 4** presents a brief knowledge about the OAs.

Depending on the levels of CV, an appropriate OA is chosen or some changes are made on the selected OA. The assignment of the CV and interaction variables to the columns is achieved by using standard linear graphs suitable for the selected OA. To determine a suitable OA for the experiment, the following procedure should be followed.

1. Determination of the number of factors and their levels

2. Determining the degree of freedom

3. Selection of OA

4. Consideration of interaction

3.8 Determination performance statistics

Defining the optimal CV requires the determination of some criteria to be optimized such as Signal / Noise ($S/_N$) ratio. The analysis of the data obtained from the experiment is made according to performance statistics and / or mean. Wrong selection of performance characteristics leads to erroneous determination of UCV levels and results. The $S/_N$ ratio is used to measure the best RD performance. Many different $S/_N$ ratios can be used depending on the purpose of the optimization process. Taguchi mentions that over than 60 $S/_N$ ratios can be used and that he developed most of them himself [2]. However, all $S/_N$ ratios must meet the criteria listed below [28].

1. The $S/_N$ ratio should reflect the variability of the UCV on the response variable.

2. The $S/_N$ ratio is independent of setting the mean. This means; the measuring system should be useful in predicting the quality even if the target value changes.

3. $S/_N$ ratio measures relative quality. Because it is used for comparative purposes.

4. $S/_N$ ratio should not cause unnecessary complexity.

Many S / N ratios are available. The three commonly used are as below.

- Largest - Best

- Smallest - Best

- Nominal - Best

3.9 Establishing the experiment and recording the results

The design optimization experiment can be done in two ways.

- Physical performance of the experiment,

- Computer simulation.

In both experiments, any combination of CV is tested for all combinations of UCV and the results are recorded. The order in which the experiments are performed should be random, as the process will not be constantly stationary. In order for the test results to be evaluated completely and precisely, the test conditions must be recorded.

3.10 Analysis of data and selection of the best values of CV

One of the goals of design optimization experiments is to reduce variability. Another goal is to adjust the mean to the target value. To achieve these two objectives, mean and performance statistics are calculated for each combination of CV in the design model. In order to evaluate the effects of CV on performance statistics and / or mean, Analysis of Variance (ANOVA) is made and percentage contributions are determined. Thus CV can be divided into three classes.

1. CV, which has a significant impact on performance statistics,

2. Setting variables that have a significant effect on average but have no effect on performance statistics,

3. Residual variables that do not affect the average or performance statistics at all.

Analysis results are plotted according to the levels of CV, so that the effects are displayed visually. The optimization procedure is different. If the performance statistics are Nominal - Best, TM uses the following two-step procedure.

1. Investigation of CVs and levels for which the analyst expects the least variability, using calculated performance statistics.

2. Investigation of the setting variables that will bring the sample mean to the target using the calculated sample mean or sample total.

With this method, the variability is reduced in the first step and the sensitivity increases in the second step. If the performance statistics are the smallest best, the TM uses a one-step procedure. This procedure aims to reduce the total variance using the calculated performance statistics; CV affecting the total variance is investigated. Levels of CV where the analyst expects the smallest mean square variability are determined. If the performance statistics are the greatest best, the TM uses a two-sided transformation. Performance statistics change from smallest to best, using the one-step method to reduce the total variance. In case of disagreement between different performance characteristics, one may be abandoned and then the best values selected. If the chosen CV combination is not included in the experiment, the performance values and confidence intervals of the best combination are estimated.

3.10.1 Analysis of variance (ANOVA)

As we mentioned earlier, DOE is used to develop or improve products or processes. The data obtained from the experiment should be analyzed. Variance analysis is used to interpret experimental data. Variance analysis was used for the first time by the British statistician Fisher. Experts usually work with samples. Because it is sometimes impossible to work with the whole population and sometimes it is very expensive. It should not be forgotten that; each individual case study forms part of the error. Sample statistics and assumptions allow the testing of hypotheses regarding experimental parameters. In order to analyze variance with sample data, we have four basic assumptions.

1. Samples are random,

2. Population is distributed normally,

3. Population variances are equal,

4. The choice of samples is independent of the others.

Total variance can be divided into two components such as inter-group variability and intra-group variability. The components of the model are tried to be estimated using the least squares method on the sample data. Total squares are used to show piecemeal variability. After calculating the total squares and determining the appropriate degrees of freedom for each component of variability, the hypothesis is tested using the F distribution [38]. A typical ANOVA table is as **Table 5**.

where

SSF = sum of squares of factor

SSe = sum of squares of error

SST = sum of squares of total

L = number of levels of the factor

N = total number of observations

F_{comp} = computed value of F

VF = variance of factor

Ve = variance of error.

After the experiment is set up, the ANOVA is completed, and the important factors and/or interactions are determined, some comments have already been made. However, if it will not be too expensive, it will be beneficial for the experimenter to learn the rest of the information. Here, we will talk about determining contribution percentages.

The rate of variability for each important factor and interaction observed in the experiment is reflected by the percentage of contribution. Percentage contribution is a function of the sum of squares of each significant factor. Percentage of contribution indicates the strength of factors and/or interaction in reducing variability. If the factor and/or interaction levels are fully controllable, the total variability can be reduced by the percentage of contribution. We know that the variance for a factor or interaction includes error variance. So we can arrange the variance for each factor to show the error variance as well.

The percentage contribution of the error provides an estimate of the adequacy of the experiment. If the error contribution percentage is 15% or less, it is assumed that no significant factor has been overlooked in the experiment. If the error contribution percentage is 50% or more, it is considered that the experimental conditions in which some important factors are ignored cannot be fully controlled or the measurement error is made [39].

In order to learn percantage contribution of the factors Pareto ANOVA can be used. Pareto ANOVA is a simplified ANOVA technique based on the Pareto principle. The Pareto ANOVA technique is a quick and easy method to analyze results of the parameter design and it does not need F-test. Pareto ANOVA does not use an F-test, but it identifies the important parameters and determines the percent contribution of each parameter [40, 21].

3.10.2 $S/_N$ ratio

Taguchi uses the statistical performance measure known as the $S/_N$ ratio used in electrical control theory to analyze the results [25]. $S/_N$ ratio is a performance criterion developed by Taguchi to select the best levels of CV that minimize the

Sources of variation	Sum of squares (SS)	Degrees of freedom	Mean square variance	F_{comp}
Factor	SS_F	L-1	$V_F = \frac{SS_F}{L-1}$	$\frac{V_F}{V_e}$
Error	SS_e	N-L	$V_e = \frac{SS_e}{N-L}$	
Total	SS_T	N-1		

Table 5.
ANOVA table.

impact of UCV [41]. The S/N ratio takes into account both mean and variability. In its simplest form, the S/N ratio is the ratio of the mean (signal) to standard deviation (noise) [4]. TM uses S/N ratios for two main purposes. The first pupose is to use the S/N ratio in order to identify CVs that reduce variability and the second purpose is to identify CVs that move the mean to target. Different S/N ratios can be choosen depending on the goal of experiment. In all cases, the S/N ratio should be maximized. Although Taguchi mentions over than 60 S/N ratios three of them such as smaller-best, larger-best and nominal-best are used frequently. The formulas of them are follows.

- Smaller-Best

$$\frac{S}{N} = -10 * \log\left(\frac{1}{n}\sum_{i=1}^{n} y_i^2\right) \tag{1}$$

- Nominal-Best

$$\frac{S}{N} = 10 * \log\left(\frac{\bar{y}}{s^2}\right) \tag{2}$$

- Larger-Best

$$\frac{S}{N} = -10 * \log\left(\frac{1}{n}\sum_{i=1}^{n} \frac{1}{y_i^2}\right) \tag{3}$$

where $\bar{y}$ is the mean of observed data, s^2 is the variance of y, n is the number of observed data, and y_i is the ith observed data.

4. Experiment

In this section, a summary of Hamzaçebi [21] is given. Hamzaçebi [21] applied the TM to determine the effects of production factors such as adhesive ratio, press pressure, and pressing time on the thermal conductivity (TC) of oriented strand board (OSB). MINITAB 17 statistical software (State College, PA, USA) was used to analyze experiments in the Taguchi design.

4.1 Data

In the article of Ref [21], adhesive ratio, pressing time, and press pressure were considered as controllable factors. **Table 6** indicates the process parameters and their levels. As deduced from **Table 6**, there are 3 factors, which have 3 levels. After the factor definitions, suitable Taguchi orthogonal array was selected as L9. The L9 design sheet and output of each experiment was given in **Table** 7.

Factors	Level 1	Level 2	Level 3
Adhesive Ratio (%)	3%	4.5%	6%
Pressing Time (minute)	3	5	7
Press Pressure (kg/cm^2)	35	40	45

Table 6.
The process parameters and their levels.

Experiment	Factors			Response	
	Adhesive ratio	Pressing time	Press pressure	$\bar{y}$	s
1	1	1	1	0.129	0.010
2	1	2	2	0.153	0.028
3	1	3	3	0.152	0.025
4	2	1	2	0.142	0.023
5	2	2	3	0.143	0.026
6	2	3	1	0.146	0.025
7	3	1	3	0.163	0.027
8	3	2	1	0.154	0.018
9	3	3	2	0.170	0.019

Table 7.
The design sheet and output of each experiment.

In **Table 7**, $\bar{y}$ and s present the mean and standard deviation of the TC values, respectively.

4.2 Solution and results

Hamzaçebi [21] was used the S/N ratio and Pareto ANOVA analysis to evaluate the results of the experiment.

Figure 3 is the main effect graph of S/N ratios that states the optimal level of the factors. The biggest S/N ratio indicated the optimal combination of parameter values. The ranking of the process parameters was obtained from S/N ratio table which is given in **Table 8**. This order was determined by comparison of delta values. The delta value is equal to the difference between maximum and minimum values for levels of each factor. **Table 8** shows that the order of importance in minimizing the TC of OSB is adhesive ratio, press pressure, and pressing time. **Figure 3** shows the optimal level of the process parameters. As deduced from **Figure 3**, the second level of adhesive ratio (3%), the first level of pressing time (3 min), and the first

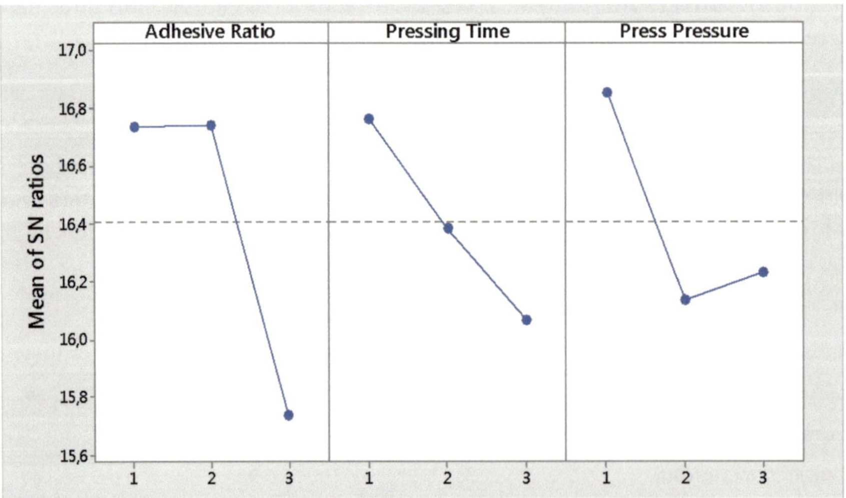

Figure 3.
Main effect plots for S/N ratios of process parameters.

Level	Adhesive ratio	Pressing time	Press pressure
1	16.74	16.77	16.86
2	16.74	16.39	16.14
3	15.74	16.07	16.23
Delta	1.00	0.70	0.72
Rank	1	3	2

Table 8.
S/N ratio values of TC.

level of press pressure (35 kg/cm^2) were the optimal values for the minimization of the TC of OSB.

Hamzaçebi [21] applied the Pareto ANOVA to determine the percent contribution of each parameter on the TC. To obtain the Pareto ANOVA of S/N ratio values, the overall mean of S/N ratios and the sum of squares due to variation about overall mean were calculated by Eqs. (4) and (5), respectively.

$$\overline{S/N} = \frac{1}{m}\sum_{i=1}^{m}(S/N)_i \tag{4}$$

where $\overline{S/N}$ is the overall mean of S/N ratio, $(S/N)_i$ is the S/N ratio for ith parameter, and m is the number of S/N ratios.

$$SS_{Total} = \sum_{i=1}^{m}\left((S/N)_i - \left(\overline{S/N}\right)\right)^2 \tag{5}$$

where SS_{Total} is the total sum of squares. Secondly, for the ith process parameter, the sum of squares due to variation about overall mean was calculated by Eq. (6).

$$SS_i = \sum_{j=1}^{k_i}\left((S/N)_{ij} - \left(\overline{S/N}\right)\right)^2 \tag{6}$$

where SS_i is the sum of the square for ith parameter, $(S/N)_{ij}$ is the S/N ratio of ith parameter of jth level, and k_im$_i$ is the number of levels of ith parameter. Finally, the contribution (Cont) of ith parameter was calculated by Eq. (7). **Table 9** presents the contribution results.

$$Cont_i = \frac{SS_i}{SS_{Total}}x100 \tag{7}$$

Process parameter	Sum of squares (SSi)	% Contribution	Rank
Adhesive ratio	0.6667	54.64	1
Press pressure	0.2456	20.13	3
Pressing time	0.3078	25.23	2
Total	1.2201	100	

Table 9.
Contribution of process parameters based on Pareto ANOVA.

4.3 Discussion

When S/N ratio results are interpreted, as it can be seen by **Table 8**, the order of importance in minimizing the TC of OSB is adhesive ratio, press pressure, and pressing time. Also, as deduced from **Figure 3**, the second level of adhesive ratio (3%), the first level of pressing time (3 min), and the first level of press pressure (35 kg/cm^2) were the optimal values for the minimization of the TC of OSB. Beside, according to the Pareto ANOVA results, the most effective factor is the adhesive ratio, the second factor is pressing time and the third one is the press pressure. The results show that the adhesive ratio is the most effective factor on the TC of OSB.

On the other hand, for this problem, if the full factorial design was used instead of using the TM, it would be necessary to set up 27 experimental setups. However, 9 experimental setups are sufficient with the TM. 18 more experimental setups are no longer required. Considering that these experiments must be repetitive, the time and cost savings gained will be appreciated. In addition, the S/N ratio criterion used to interpret the results facilitates the decision made by the decision maker.

5. Conclusion

The objective of this study is to give a brief knowledge about the TM which is used in both manufacturing and service sectors as an optimization tool for product and process. Literature of the TM applications is very large and it is stil growing. The objective of the TM is to setup a RD, hence reduce the variabilty of performance characteristcs of the product and/or process. The main advantage of the TM is cost reduction in time and budget.

In order to present an example, the summary of Hamzaçebi [21] is given. The theoretical benefits of the TM can be seen as follows from the result of Ref. [21].

1. TM is a powerful technique to analyze the effects of the process parameters.

2. Time and cost of experiments can be reduced by using TM. As a result of a selected orthogonal array, 9 experiments were performed instead of 27 experiments, which should be done for full factorial design implementation.

3. The same results were obtained by both S/N ratio analysis and Pareto ANOVA. Thus, it can be said that the outputs of the analysis is consistent.

Author details

Coşkun Hamzaçebi
Karadeniz Technical University, Trabzon, Turkey

*Address all correspondence to: chcebi@gmail.com

References

[1] TaguchiG., Chowdhury S., Wu Y., Taguchi's Quality Engineering Handbook, John Wiley&Sons, 2005, NJ

[2] Pignatiello, J. J., An Overview of The Strategy And Tactics of Taguchi, IIE Transactions, 20, 1988

[3] Clausing, D., Taquchi G., Robust Quality, MA: Harvard Business Review, Jenuary-February 1990

[4] Unal, R., Dean, E. B., Taguchi Approach To Design Optimization For Quality And Cost: An Owerview, Annual Conferennce Of The International Society Of Parametric Analysts, 1991

[5] Antony J., Design of Experiments for Engineers and Scientists, Elsevier, 2014, Scotland,UK

[6] Sharma V., Misrab J. P., Singhala P., Optimization of process parameters on Combustor Material Using Taguchi & MCDM Method in Electro-Discharge Machining (EDM), Materials Today: Proceedings, 18, 2019

[7] Kumar D., Mondal S., Process parameters optimization of AISI M2 steel in EDM using Taguchi based TOPSIS and GRA, Materials Today: Proceedings, 26(2), 2020

[8] Viswanathan R.,.Ramesh S. , Maniraj S., Subburam V.,Measurement and multi-response optimization of turning parameters for magnesium alloy using hybrid combination of Taguchi-GRA-PCA, Measurement, 159, 2020

[9] Liu S.J., C.-H. Lin, and Y.-C. Wu, "Minimizing the sinkmarks in injection-molded thermoplastics," Advances in Polymer Technology, 20(3), 2001

[10] Lan T.S., M.-C. Chiu, and L.-J. Yeh, "An approach to rib design of injection molded product using finite element and Taguchi method," Information Technology Journal, 7(2), 2008.

[11] Asafa T. B., Tabet N., Said S.A.M, Taguchi method–ANN integration for predictive model of intrinsic stress in hydrogenated amorphous silicon film deposited by plasma enhanced chemical vapour deposition, Neurocomputing, 106, April, 2013

[12] Parinam S., Kumar M., Kumari N., Karar V., Sharma A.L, An improved optical parameter optimisation approach using Taguchi and genetic algorithm for high transmission optical filter design, Optik, 182, April 2019

[13] Sekulic M., Kovac P., Gostimirovic, M., Kramar, D., "Optimization of high-pressure jet assisted turning process by Taguchi method", Advances in Production Engineering & Management, 8(1), 2013

[14] Fei N. C., Mehat N. M., Kamaruddin S., "Practical Applications of Taguchi Method for Optimization of Processing Parameters for Plastic Injection Moulding: A Retrospective Review", , ISRN Industrial Engineering, 2013

[15] Dave S. V., Bhogayata A., "The strength oriented mix design for geopolymer concrete using Taguchi method and Indian concrete mix design code", Construction and Building Materials, 262, November, 2020

[16] Terzioğlu H., "Analysis of effect factors on thermoelectric generator using Taguchi method", Measurement, 149, January 2020

[17] Zhou K., Mao J., Li Y, Xiang J., "Parameters optimization of borehole and internal thermal resistance for single U-tube ground heat exchangers using Taguchi method", Energy Conversion and Management, 201, December 2019

[18] Hong Chien-Wen, "Using the Taguchi method for effective market segmentation", Expert Systems with Applications, 39(5), 2012

[19] Kumar V., Kharb R., Chaudhary H., "Optimization & design of isradipine loaded solid lipid nanobioparticles using rutin by Taguchi methodology", International Journal of Biological Macromolecules, 92, November 2016

[20] Tiryaki, S., Hamzaçebi, C., and Malkoçoğlu, A., Evaluation of process parameters for lower surface roughness in wood machining by using Taguchi design methodology, European Journal of Wood and Wood Products, 73, 2015

[21] Hamzaçebi C., "Optimization of Process Parameters in Oriented Strand Board Manufacturing by Taguchi Method", BioResource, 11(3), 2016

[22] Alafaghani A., Qattawi A., Investigating the effect of fused deposition modeling processing parameters using Taguchi design of experiment method, Journal of Manufacturing Processes, 36, December 2018

[23] Mitra Anirban C., Jawarkar M, Soni T., Kiranchand G. R., Implementation of Taguchi Method for Robust Suspension Design, Procedia Engineering, 144, 2016

[24] Çakıroğlu R., Acır A., Optimization of cutting parameters on drill bit temperature in drilling by Taguchi method, Measurement, 46(9), 2013

[25] Phadke, S. M., , Introduction To Quality Engineering, Asian Productivity Organizatio, 1989, Dearborn

[26] Hamzaçebi C., Kalite Yönetiminde Taguchi Felsefesi, 2000, Gazi Üniversitesi, Ankara, Türkiye

[27] Tsui, K. L., An Overview of Taguchi Method And Newly Developed Statistical Methods For Robust Design, IIE Transactions, 24, 1992

[28] Connor, A.M., Parameter Sizing For Fluid Power Circuits Using Taguchi Methods, Journal Of Engineering Design, 10(4),1999

[29] Sudhakar, P. R., An Introduction To Quality Improvement Thtough Taguchi Methods, Industrial Engineering, January, 1995,

[30] Şirvancı M., Kalite İçin Deney Tasarımı, Literatür, 1997,İstanbul

[31] Besterfield, D. H., Besterfield,C., Besterfield, G. H., Besterfield, M., Total Quality Management, Prentice Hall Inc., 1995,New Jersey

[32] Çelik, C, Burnak, N., Asystematic Approach To The Solution Of The Design Optimization Problem, Total Quality Management, 9,1998,

[33] Ross, P. J., Taguchi Techniques For Quality Engineering, Mc Graw Hill, 1988, Newyork

[34] Kackar, R., "Off-Line Quality Control, Parameter Design And The Taguchi Method", Journal Of Quality Technology, 17, 1985

[35] Taguchi, G., Wu, Y., , Introduction To Off-Line Quality Control,Central Japan Quality Control Association, 1979, Nagaya

[36] Taguchi, G., System of Experimental Design, ASI, 1991, Dearborn

[37] Unal, R., Dean, E. B., ,Design For Cost And Quality: The Robust Design Approach, Journal Of Parametrics, 11 (1), 1991

[38] Sower, V. E., Savoıe, M. J., Renick, S., An Introduction to Quality Management And Engineering, Prentice Hall Inc., New Jersey, 1999

[39] Krishnaiah K., Shahabudeen P., Applied Design of Experiments and Taguchi Methods, PHI Learning Private Limited, 2012

[40] Venkateswarlu, G., Davidson, M. J., and Tagore, G. R. N. Influence of process parameters on the cup drawing of aluminum 7075 sheet, International Journal of Engineering Science 2, 2010

[41] Bryne, D. M. Ve Taguchi, S., The Taguchi Approach To Parameter Design, ASQC Quality Congress Transactions, Anaheim, 1986

Chapter 8

Optimized Portfolios: All Seasons Strategy

Raúl D. Navas, Sónia R. Bentes and Helena V.G. Navas

Abstract

Our study explores the efficient frontier of optimal investment, taking behind the Markowitz's theory, while advocating a diversified portfolio to reduce risk. To perform it, six portfolio models are proposed, and its formation are made by a solver, where the selected solving method is the GRG Nonlinear engine for linear solver problems. Our main goal is to design portfolios that resists to financial crisis but at the same time persists in a wealthy period. We analyze the decade where we assisted to two crashes (2000–2010) and a semi-decade where we assist to a wealthy period (2011–2018). The assets used are varied, such as Equities indexes form various countries, sector equities, bonds, commodities, EURUSD exchange and VIX. Results show that the GRG Nonlinear engine is powerful, providing excess returns in all six models.

Keywords: MPT, Markowitz, portfolios' formation, sharp-ratio, volatility

1. Introduction

The inspiration for our work comes from the well-known investor, Ray Dalio who built a considerable personal fortune with the incredible success of the Pure Alpha strategy. In the mid-1990s he began to think about his inheritance and funds he wanted to leave behind and asked this question: "What kind of portfolio would you use if you were not already present to actively manage money?" What kind of portfolio would survive your own decision-making and would continue to support their children and their philanthropic efforts for decades? [1].

A brand-new look at asset placement. A new set of rules. And only after the portfolio has been retrospectively tested until 1925; only after having produced consistent results in a variety of economic conditions, Ray Dalio began to offer it to a narrow group. The new strategy, known as the "All Seasons" strategy, was publicly unveiled in 1996, just four years before a mass market correction put it to the test. "Passed" with distinction [1].

Conventional wisdom and the conventional management of a portfolio leave us in the hands of a model that continually shows that it cannot survive when times are tough. So, we began to explore whether we could define portfolios - asset distribution - that would perform well in any economic environment in the future, such as in the year 2008, a depression or a recession. Because no one knows what is going to happen in five years, how much more in 20 or 30 years.

According to [1], having into account this basis, we propose six different models, aiming to maximize returns but at the same time, reduce risk. Theory behind this is the Harry Markowitz's [2, 3], who is known as the father of modern portfolio theory. It explains in this way and synthesized the fundamental concept behind the work that earned him the Nobel Prize: investments in a portfolio should not be seen individually, but as a group. There is a trade-off between risk and return, so "do not listen to just one instrument, listen to the entire orchestra". How investments behave together and how they are diversified will determine return. This advice may seem simple now, but in 1952 this thought was groundbreaking. Somehow this approach influenced virtually all portfolio managers from New York and Hong Kong.

We combine portfolios with a wide range of equity (mainly indexes from various countries and the main sectors as well), different kind of bonds (US and German treasury bonds and corporate bonds), a range of commodities (for example, different metals, agriculture commodities, energy commodities, etc.), EUR/USD exchange and VIX (volatility index of S&P 500) through a solver using the GRG (Generalized Reduced Gradient) Nonlinear engine for linear solver problems. The spam range is from 2000 to 2018 in order to cover two market crashes (2002 – technological and 2008 – subprime) and a good decade forward. Our mains investigation question is if it is possible to create a portfolio or a set of portfolios that presents robust results in a bad decade but, at the same time, in a good decade as well? Results show that definitively, is doable.

Next section, literature review, we explore the theory behind the concept of this study and empirical achievements from different authors. Section 3 presents six different models where we are going to using the solver, Section 4 preliminary analysis to the data set, in Section 5 we present the results of the models and we propose some portfolios to use as well. Finally, Section 6 concludes.

2. Literature review

Many investors are naive in their financial beliefs and do not understand basic concepts such as equity or diversification [4, 5]. Benjamin Graham 1949 *apud* [6], the father of the value investing, proposed that an equilibrated portfolio should be constituted by 50% equity and 50% bonds; an intelligent investor may own 100% equity in his portfolio in certain conditions, the most important of them: only if in a crash crisis, the portfolio presented a positive return. By dividing the money by 50% for stocks and 50% for bonds (or some similar variation), many investors would think they were diversifying and reducing their risk. But later, when [2] presented his work about the efficient portfolio, concluded that what investors are doing is taking more risks than they think. Because, according to Ray Dalio *apud* [1–3] shares can be three times riskier (i.e., volatile) than bonds. In fact, by having a 50/50 portfolio, we have something more like a 95% risk distribution in stocks. Below, **Figure 1** represents a chart with a 50/50 portfolio. The left side shows the money divided by shares and bonds, in percentage. The right side shows the same portfolio, but divided in terms of risk, between stocks and bonds.

At first glance, with 50% of the money in shares, it seems relatively balanced. But, as it turns out here it would have been about 95% risk, given the volatility of its composition in stocks. So, if shares sink, the whole portfolio sinks. And the balance is lost. How does this concept work into real life? From 1973 to 2013, the S&P 500 lost money nine times and accumulated losses totaled 134%. In the same period, bonds (represented by the Barclays Aggregate Bond index) lost money only three times and accumulated losses were 6%. Therefore, having a 50/50 portfolio, the S&P 500 would have accrued 95% of the losses.

Placing assets is the only key that can differentiate us from all investors [2, 3]. Nobel Prize winner and father of modern portfolio theory (MPT) said that "diversification is the only free lunch." Why? Because spreading the money for different investments lowers the risk and increases the possibility of gains over time and costs nothing.

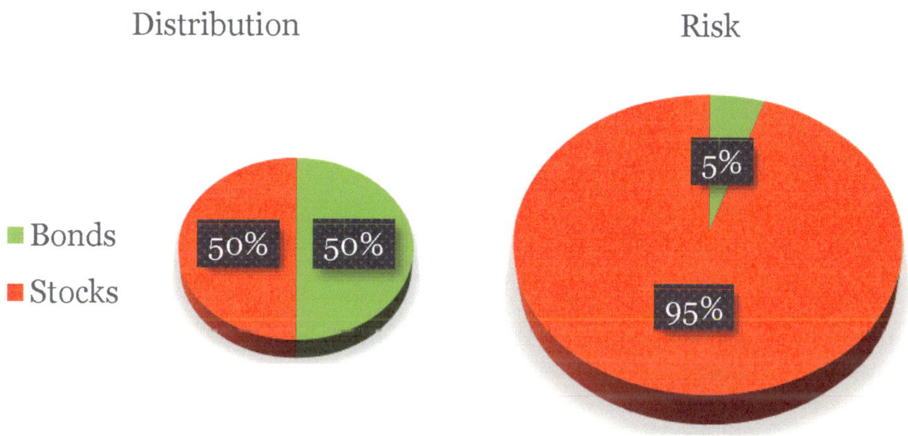

Figure 1.
Allocation versus risk. Source: adapted from [1].

	GROWTH	INFLATION
RISE ↑	Higher than expected economic growth	Higher than expected inflation
FALL ↓	Lower than expected economic growth	Lower than expected inflation

Table 1.
The four things that drive asset prices. Source: adapted from [1].

When we look at most portfolios, they usually hold up well in good periods, but they fall in bad periods. And then, the strategy is simply to wait for the stock to go up. This conventional approach to diversifying investments is not at all diversified (Dalio *apud* [1]). According to [7] "Financial crises occur in all market economies, although sometimes there are long periods of quiet. Crises occur in developed countries, not just emerging markets. Crises occur in economies with and without a central bank and with and without deposit insurance".

Competitive pressures and market efficiency turn difficult to financial forecast - particularly to predict asset returns - is very difficult compared to standard forecasting problems in macroeconomics, in which the presence of a sizeable persistent component makes forecasting easier [8].

Dalio *apud* [1] revealed the simplest and most important distinction of all. There are only four things that drive asset prices (**Table 1**):

1. Inflation

2. Deflation

3. High economic growth

4. Declining economic growth

In this way, it all boils down to four possible environments, or economic "seasons," which will ultimately determine whether investments (asset prices) rise or fall - except that, unlike the seasons, there is no predetermined order in the succession. They are:

1. Inflation higher than expected (rising prices)

2. Inflation lower than expected (deflation)

3. Economic growth higher than expected

4. Economic growth lower than expected

The price of a stock (or a bond) already incorporates what we (the market) "expect" about the future. Many authors [9–11], claim that there is literally a picture of the future when looking at prices today. In other words, the stock price of a company today already incorporates the expectations of investors, who believe that the company will continue to grow at a certain pace [12–14] – this phenomenon also known as efficient market hypothesis (EMH). This is why is sometimes heard that the stock price will fall when companies announce that their future growth (their profits) will be lower than they had originally forecast – see also the post-earnings announcement drift (PEAD) phenomenon [15–17].

It is the surprises that will ultimately determine which asset class will behave well. If the news announces that there will be sustained growth, this will be very good for stocks and not so good for bonds. If we watch a surprise fall in inflation this will be good for the bonds [18, 19]. If there are only four potential economic environments, or seasons, one should therefore have 25% of the risk in each of the categories. That is why this approach is called "All Seasons" because there are four possible seasons in the financial world, and no one really knows which season will come next – EMH/Random walk [12, 20–22]. With this approach each season, each quadrant is always covered, so the portfolio is always protected. Let us imagine, then, four portfolios, each with an equivalent amount of risk. This means that we will not have exposure to any particular environment. We are not trying to predict the future, because no one knows what the future will bring [12, 22–24]. What is known is that there are only four potential seasons that we will have to face. Using this investment strategy, we can know that we are protected - not just hopeful - and that the investments are safe and will perform well in any season that comes.

"All Seasons": today we can structure a portfolio that will behave well in 2029, even if we have no chance of knowing what the world will look like in 2029. Below is a table that shows the four potential seasons and the type of investment that will perform best in each of these environments, categorizing each of them in each of the seasons (**Table 2**).

The original "All Seasons" is composed by equity, bonds and commodities which became a popular asset over the past decade [25]. [26] argues that MPT is the formula of diversification, which selects a collection of assets that has collectively lower risk than individually. In sum, for a given amount of risk, MPT describes how to select a portfolio with the highest possible expected return [27, 28]. Below it is presented, in **Figure 2**, the efficient frontier.

	GROWTH	INFLATION
RISE ↑	Equities Corporate bonds Commodities/gold	Commodities/gold TIPS
FALL ↓	Treasury bonds Treasury Inflation-Protected Securities (TIPS)	Treasury bonds Equities

Table 2.
List of assets for each "season".

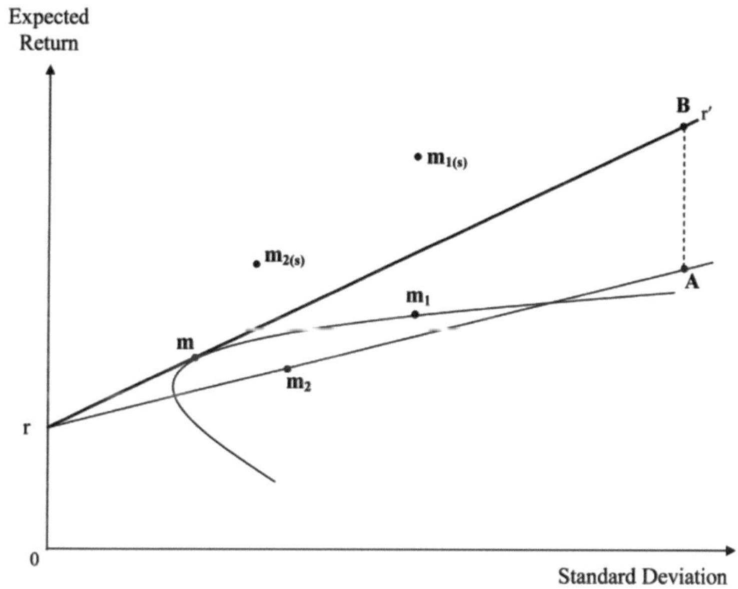

Figure 2.
Efficient frontier. Source: [29].

The hyperbola is sometimes referred to as the "Markowitz Bullet" and is the efficient frontier if no risk-free asset is available. With a risk-free asset, the straight line is the efficient frontier.

The Capital Asset Pricing Model (CAPM), for example, was the next step, it approached the risk of an individual asset through the diversification theory [30].

Based on this theory background and MPT, we present in the next chapter six different portfolios aiming to a certain risk, produce the maximum return to the investor.

3. Model framework

Six portfolio models are proposed: first, it is used a solver, where the selected solving method is the Generalized Reduced Gradient (GRG) Nonlinear engine for linear solver problems. The form is:

$$\max f(x): h(x) = 0, L \leq x \leq U, \tag{1}$$

Where h has dimension m. The method supposes can be partition $x = (v,w)$ such that:

- v has dimension m (and w has dimension n-m);
- the values of v are strictly within their bounds: $L_v < v < U_v$ (this is a nondegeneracy assumption);
- $\nabla_v h(x)$ is nonsingular at $x = (v,w)$.

As in the linear case, for any w there is a unique value, $v(w)$, such that $h(v(w),w) = 0$ (c.f., Implicit Function Theorem), which implies that

Equities index	Bonds	Commodities	Other
S&P 500	Treasury 1–3 years	Cocoa	EUR/USD
Dow Jones 30	Treasury 7–10 years	Coffee	VIX
NASDAQ	Treasury 20+ years	Corn	
EuroStoxx 600	TIPS	Sugar	
Hang Seng	Corporate bonds	Gold	
Emerging Markets	Bunds (Germany)	Copper	
Real Estate		Silver	
Consumer		Crude	
Healthcare		Natural Gas	
Communications		General commodities index	
Energy			
Financials			
Industrials			
Semiconductors			

Notes: S&P = Standard & Poor's; TIPS = Treasury Inflation Protected Securities; EUR/USD = Euro vs. USA Dollars exchange; VIX = Volatility Index (S&P 500).

Table 3.
Assets list.

$dv/dw = \left(\nabla_v h(x)\right)^{-1} \nabla_w h(x)$. The idea is to choose the direction of the independent variables to be the reduced gradient: $\nabla_w \left(f(x) - y^T h(x)\right)$, where $y = dv/dw = \left(\nabla_v h(x)\right)^{-1} \nabla_w h(x)$. Then, the step size is chosen, and a correction procedure applied to return to the surface, $h(x) = 0$.

The main steps (except the correction procedure) are the same as the reduced gradient method, changing the working set as appropriate.

The constitution of the portfolios is set from the Solver and varies. The six portfolios present different risks and returns, depending of the profile of each investor. There are conservative portfolio and aggressive portfolios. The solver configuration of each portfolio is showed above.

With regards to variable cells, the percentage of weighing of the assets type are the changeable ones. It is used a wide asset as equity indexes, bonds, commodities and other. The detail of each family of assets used in the model are listed table above. In total, it is used 32 assets:

With regards to subject to the constraints, the sum of the percentage of each asset is equal to 1, i.e., 100%:

$$\sum x \, Assets(a) = 1 \tag{2}$$

Where x = coefficient; a = each type of asset as showed in **Table 3** – Assets list.

Note: it is forced to make unconstrained variables non-negative.

The period is set between 2000 to 2018, but in some analysis the two decades are separated (2000 to 2010; and 2011–2018). The reason of the period spam used is important because:

1. The first decade (2000 to 2010) was very turbulent for financial markets, where occurred two crashes:

    a. Between 2001 and 2002 the technological crash;

    b. Between 2008 and 2009 the subprime crisis.

2. The second semi-decade not completed yet, between 2011 to 2018 where there is a recover from the last decade.

Then is important to study some robust portfolios that can provide some return to investors and at the same time with lower risk, mean, volatility, in order to be prepared for crashes or deflationary periods.

The model uses past returns (monthly returns) for each asset and the portfolios are re-equilibrated monthly according to the optimal weighting of each one. The benchmark, to compare results is the S&P 500 index. It is relied on monthly returns, computed as given by:

$$R_t = \frac{P_t}{P_{t-1}} - 1 \quad (3)$$

Where R_t = monthly returns; P_t and P_{t-1} are the assets prices at moments *t* and *t-1* respectively.

Finally, it is presented above, for each model, the own specifications and objectives of each one:

3.1 Model 1: maximize sharp ratio

The set objective of this model is to maximize the sharp ratio for all the period (2000–2018). It is relied by the division by the average year return and the standard deviation, computed as given by the following steps:

$$S = P(1+i)^n \quad (4)$$

Where S = Accumulated value; P = Principal.

That is:

$$FV = PV(1+i)^n \quad (5)$$

Where FV = Future Value; PV = Present Value; i = rate; n = number of periods (years).

But is need the rate for the numerator:

$$i = \sqrt[n]{\frac{FV}{PV}} \quad (6)$$

The, the standard deviation (σ) – the denominator – is a measure of how widely values are dispersed from the average value (the mean), using the "n" method. It is used the following formula:

$$\sigma = \sqrt{\frac{\Sigma(x-\bar{x})^2}{n}} \tag{7}$$

Finally, the Sharp Ratio (SR) formula:

$$SR = \frac{i}{\sigma} \tag{8}$$

This is a measure of stability, if SR > 1 it means that returns overcome the standard deviation (volatility).

3.2 Model 2: maximize rate return

In this model, the concern is to bring the maximum return to the investor, ignoring the volatility, then we can argue that model 2 presents the higher risk comparing to others:

Set objective: Global rate return
To: Maximum

3.3 Model 3: two decades, equals returns

The model equals the return of the two decades ($i_{2000-2010} = i_{2011-2018}$). Comparing to other models, it may generate a more distributed income to investors. Then, in a decade of crises the investor may generate the same return as in a period of expansion. Also, model 3 "guarantees" a minimum return of half of percent each year:

Set objective: Global rate return
To: Maximum
Additional subject to the constrains: $i_{2000-2010} = i_{2011-2018}$; i_n > 0,5%

3.4 Model 4: maximize rate and sharp-ratio

In this model, the concern is to bring some extra return to the investor. It may generate more income than the model 1 but still with the concern of a stability, lowering a little bit the volatility of the portfolio:

Set objective: Global rate return
To: Maximum
Additional subject to the constrains: Sharp Ratio > 1 (all model)

3.5 Model 5: maximize rate and sharp-ratio (version 2)

In this model, in a similar way of the previous model (model 4), the concern is to bring some extra return to the investor but still with the concern of a stability, lowering a little bit the volatility of the portfolio:

Set objective: Global rate return

To: Maximum

Additional subject to the constrains: SR > 1 (period 2000–2010); SR > 1 (period 2011–2018)

Comparing to the previous model, the SR >1 appears twice in the constrains and not in the whole model (2000–2018). This measure will provide stability in the first decade but in the second decade as well.

3.6 Model 6: maximize the minimal year return

The model "guarantees" a minimal return year by year. Then, it may generate positive returns each year. Basically, it maximizes the minimum:

Set objective: Minimum return of each year (2000–2018)

To: Maximum

Additional subject to the constrains: Year return > Minimum return of each year

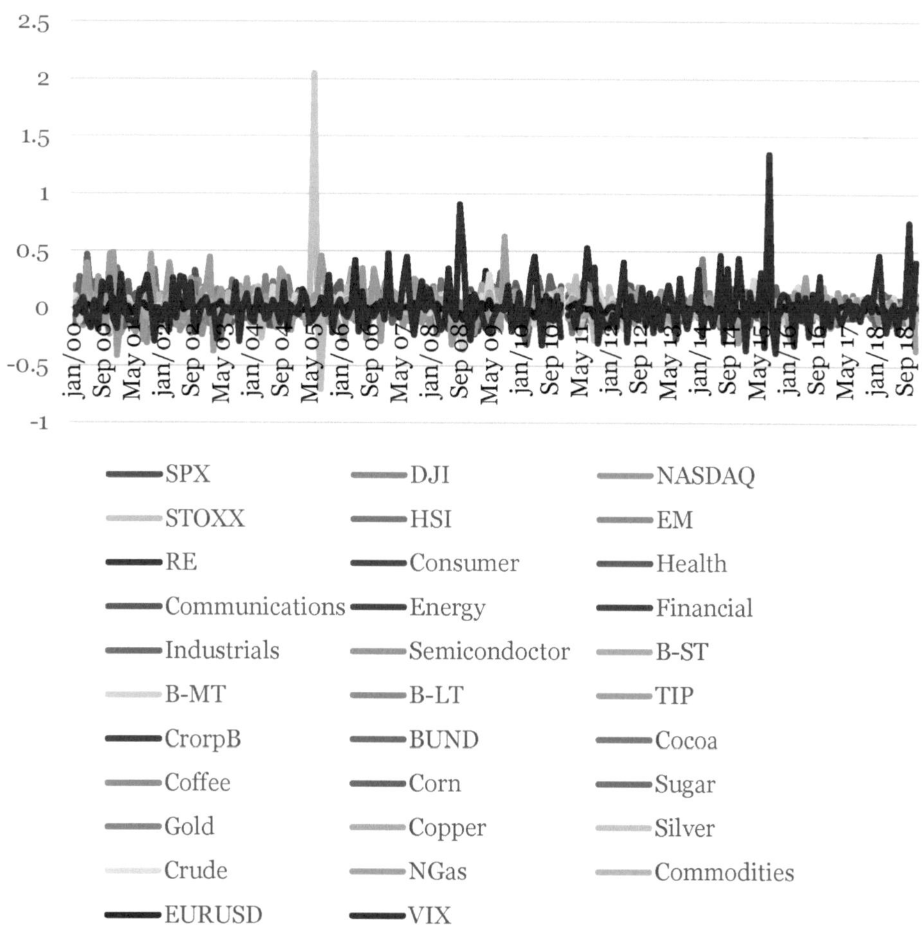

Figure 3.
Monthly returns of the 32 assets. Notes: SPX = Standard & Poor's ticker; DJI = Dow Jones Industrials ticker; STOXX = Eurostoxx 600 ticker; HSI = Hang Seng Index; EM = Emerging Markets; RE = Real Estate; B-ST = US Bonds Short Term; B-MT = US Bonds Medium Term; B-LT = US Bonds Long Term; TIP = Treasury Inflation Protected Securities; CorpB = Corporation Bonds; NGas = Natural Gas; EUR/USD = Euro vs. USA Dollars exchange; VIX = Volatility Index (S&P 500).

4. Preliminary data analysis

Market-adjusted prices data were collected from Yahoo Finance and from the Investing.com databases for all assets between 2000 and 2018. Monthly data for the assets informs the computation of returns. **Figure 3** reports the fluctuations of the months returns, illustrating the synchronized behavior of the returns compared with prices (**Figure 4**). Correlation matrix and collinearity statistic were made (table to big, then only available by request) and descriptive statistics of monthly returns of the assets in **Table 4**.

The clusters are quite evident: volatility is present during the period. It is noticed also that spikes vary in time and between the assets themselves which is expected according to the propose in this study in order to create an adequate and a stable portfolio for "four stations". In general, spikes are more evident in VIX, which means this is the asset with more variation in prices (volatility). We also can see two evident clusters in this asset during the crisis of 2008 and before October of 2015 (fears about China). It is noticed also that silver had an evident cluster after April of

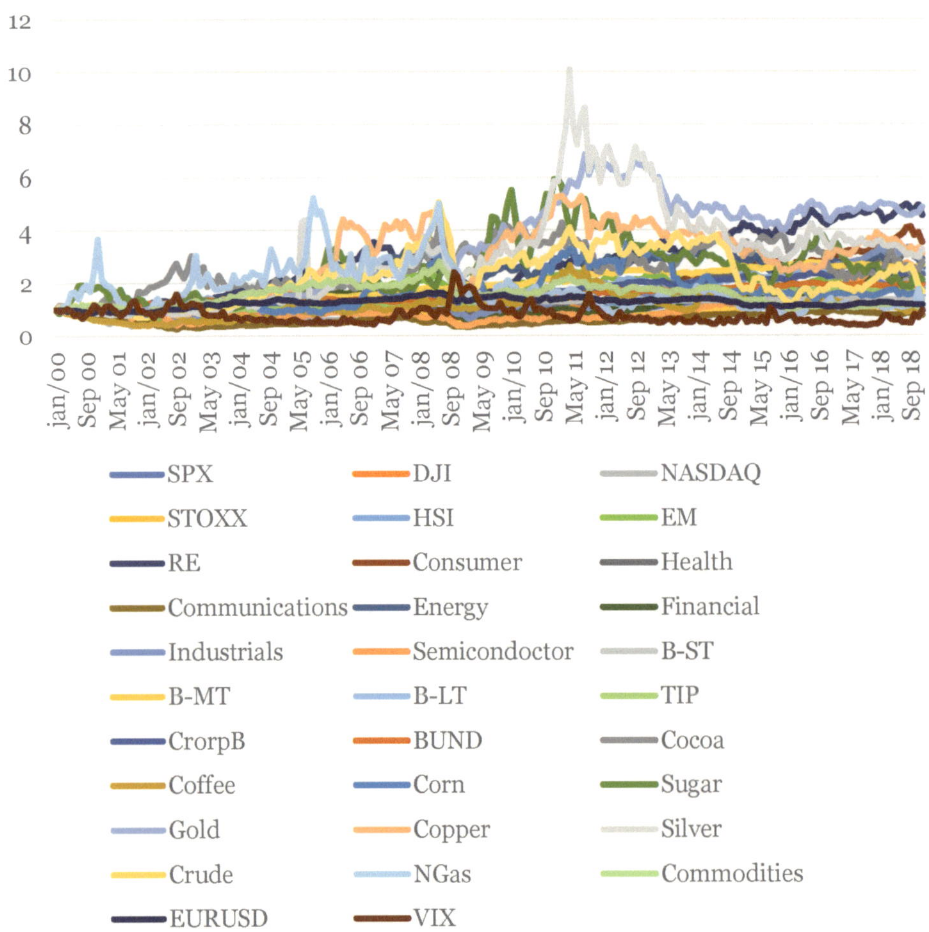

Figure 4.
Accumulated returns of the 32 assets. Notes: SPX = Standard & Poor's ticker; DJI = Dow Jones Industrials ticker; STOXX = Eurostoxx 600 ticker; HSI = Hang Seng Index; EM = Emerging Markets; RE = Real Estate; B-ST = US Bonds Short Term; B-MT = US Bonds Medium Term; B-LT = US Bonds Long Term; TIP = Treasury Inflation Protected Securities; CorpB = Corporation Bonds; NGas = Natural Gas; EUR/USD = Euro vs. USA Dollars exchange; VIX = Volatility Index (S&P 500).

	Min	Max	Mean	Standard Deviation	Var	Kurtosis
SPX	−0,16,942	0,107,723	0,003389	0,041935	,002	1208
DJI	−0,14,060	0,106,046	0,004066	0,040566	,002	1026
NASDAQ	−0,22,901	0,191,947	0,004314	0,063888	,004	1478
STOXX	−0,14,134	0,134,717	0,000475	0,043185	,002	,981
HSI	−0,19,218	0,191,929	0,003585	0,060497	,004	,506
EM	−0,25,578	0,168,629	0,005717	0,064815	,004	,757
RE	−0,30,435	0,325,359	0,008478	0,058845	,003	7569
Consumer	−0,16,678	0,135,928	0,006671	0,046783	,002	1075
Health	−0,14,249	0,085279	0,005354	0,035844	,001	1247
Communications	−0,19,721	0,323,605	0,000458	0,059830	,004	4326
Energy	−0,17,820	0,169,379	0,005666	0,055412	,003	,865
Financial	−0,22,757	0,217,848	0,004584	0,054803	,003	3504
Industrials	−0,19,954	0,194,630	0,005873	0,052299	,003	1912
Semicondoctor	−0,30,345	0,238,355	0,006824	0,079641	,006	1485
B-ST	−0,01401	0,023438	0,001383	0,003911	,000	6893
B-MT	−0,05473	0,077308	0,004492	0,017663	,000	1557
B-LT	−0,13,070	0,138,855	0,004450	0,035025	,001	3182
TIP	−0,08111	0,065035	0,002477	0,015140	,000	6606
CrorpB	−0,10,723	0,133,314	0,003885	0,019498	,000	12,713
BUND	−0,03512	0,047787	0,003007	0,015447	,000	-,203
Cocoa	−0,28,082	0,345,646	0,008779	0,093491	,009	,874
Coffee	−0,22,600	0,436,102	0,003457	0,090881	,008	2374
Corn	−0,26,536	0,221,904	0,005798	0,085273	,007	,433
Sugar	−0,31,247	0,463,178	0,008467	0,103,947	,011	1641
Gold	−0,18,005	0,138,671	0,008176	0,048522	,002	,752
Copper	−0,36,149	0,340,836	0,007984	0,076578	,006	3505
Silver	−0,70,670	2,047,420	0,015425	0,168,140	,028	96,160
Crude	−0,32,621	0,297,143	0,006479	0,092205	,009	,572
NGas	−0,41,616	0,626,133	0,012380	0,157,891	,025	1771
Commodities	−0,22,325	0,137,865	0,001168	0,047982	,002	1895
EURUSD	−0,09720	0,101,047	0,001066	0,029126	,001	1183
VIX	−0,38,489	1,345,709	0,019859	0,217,513	,047	6695

Notes: Min = Minimum; Max = Maximum; Var = Variance; SPX = Standard & Poor's ticker; DJI = Dow Jones Industrials ticker; STOXX = Eurostoxx 600 ticker; HSI = Hang Seng Index; EM = Emerging Markets; RE = Real Estate; B-ST = US Bonds Short Term; B-MT = US Bonds Medium Term; B-LT = US Bonds Long Term; TIP = Treasury Inflation Protected Securities; CorpB = Corporation Bonds; NGas = Natural Gas; EUR/USD = Euro vs. USA Dollars exchange; VIX = Volatility Index (S&P 500).

Table 4.
Descriptive statistics of monthly returns of the 32 assets.

2005 when other assets remained stable. What regards to equity, the most positive cluster (i.e. low spike) is present after May of 2009 when the market was recovering from the crisis.

If compared to the next figure (**Figure 4**), it is illustrated the synchronized behavior of the returns compared with prices. The spikes are much more evident. It also offers a clear picture of the volatility clusters.

What refers to correlation matrix and collinearity statistic (table available by request), US markets are very correlated and collineated with European market and all equity-sectors, although there is no correlation with the Chinese market, bonds, commodities and EUR/USD exchange. Also, it is shown that VIX is inverse correlated to the equity market in general. What regards to bonds, there is a high correlation and collinearity between themselves (except for the short-term bonds which are only a little correlated to medium-term bonds) but there is not (in general) with commodities and VIX. Commodities, in general, are not correlated with themselves (except crude oil that is correlated to all commodities index) and not correlated either to VIX or EUR/USD exchange. It is interesting to note that agriculture commodities are not correlated at all to themselves (cocoa, coffee, corn or sugar) but neither metals, for example, are not correlated between themselves (gold, copper or silver).

As the table above shows, standard deviation presents higher values rather the mean which means that volatility is present for all types of assets. Also, kurtosis presents value higher than 3 for Real State (equity), communications, financial services, short-term bonds, long-term bonds, TIPs, corporate bonds, copper, VIX and an exceptional high value (higher than 96) for silver. This may mean that the monthly returns distribution is non-normal for this kind of assets.

5. Results

This study uses GRG Nonlinear engine for linear solver problems. **Table 5** reports the returns from each portfolio (Model 1 – Model 6) and **Table 6**, the constitution of assets for each portfolio.

Rate means the yearly return of the portfolio, and as can be seen the best result of 15,06% belongs to portfolio 2 which is expected because we are maximizing this metric (rate), although in a less consistent way since sharp ratio presents the lowest value compared to other portfolios. This means that portfolio 2 is the most volatile, i.e., in terms of sharp ratio almost equals to the benchmark (S&P 500). Still it has fewer negative years when comparing to the benchmark (3 versus 7). Although, it loses power in the good semi-decade (2011–2018), showing a return of 8,55% (annually), when in the bad decade (2000–2010) the average return was 20,03%. Portfolio 1 presents the highest sharp ratio with no negative years; the worst year presented a positive return of 2,18%. The average yearly return is 4,49% in the overall and its maximum presented a value of 6,52% (much lower comparing to 29,60% - the benchmark). It means that this portfolio is adequate to a very conservative investor. The rate is only a little higher than a deposit rate, which is expected according to its constitution (see **Table 6**) because 65% is constituted by treasury bonds, then only 16% equity, 7% commodities and 12% others (EUR/USD and VIX). In Portfolio 3 we try to create a portfolio that, in general, the rate of a bad decade is almost equal to a good decade. It is expected a yearly rate of 14,15% overall and equal for both decades. Sharp ratio still present positive values (superior to 1) and the investor should not present any negative years with Portfolio 3 (there was this condition as well in this model). To accomplish that, portfolio constitution is curious: 42% equity, 32% VIX, 26% commodities and no bonds (see **Table 6**). In Portfolio 4 the overall rate is maximized but with the constrain of the overall sharp ratio equal or superior to 1 and the solver obtained the result successfully. The overall rate is 14,83% by year which is an excellent result, but it loses "power" in the good decade (18,28% 2000–2010 vs. 10,25% 2011–2018). The worse year was in 2013 (−10,14%)

Panel A: Decade 2000–2010							
	SP500 Benchmark	**Model 1 Max SR**	**Model 2 Max return**	**Model 3 Equal return**	**Model 4 Max return/SR**	**Model 5 Model 4 (ver2)**	**Model 6 Max min**
2000	−5,32%	4,96%	29,62%	9,18%	19,89%	20,50%	8,75%
2001	−13,04%	4,02%	−13,37%	0,50%	−4,00%	−8,10%	11,23%
2002	−23,37%	4,79%	27,08%	6,96%	20,85%	12,26%	12,82%
2003	26,38%	5,31%	10,43%	12,23%	11,57%	19,94%	13,20%
2004	8,99%	4,54%	4,65%	0,50%	6,69%	1,79%	2,59%
2005	3,00%	4,21%	73,67%	37,70%	54,60%	59,50%	27,36%
2006	13,62%	3,96%	17,76%	16,80%	21,49%	16,33%	15,64%
2007	3,53%	6,08%	39,86%	32,98%	31,60%	32,30%	26,42%
2008	−38,49%	3,61%	13,02%	10,51%	8,95%	2,49%	7,33%
2009	23,45%	5,20%	9,57%	10,67%	12,29%	17,95%	18,50%
2010	12,78%	6,17%	27,57%	23,86%	26,61%	26,33%	17,49%
Rate	−0,93%	4,80%	20,03%	14,15%	18,28%	17,12%	14,44%
ER	0	5,74%	20,97%	15,09%	19,22%	18,06%	15,37%
SR	−0,05	5,97	0,94	1,22	1,24	1,00	2,00
Panel B: Semi-decade 2011–2018							
2011	−2,22%	4,76%	11,76%	14,59%	12,16%	9,32%	2,58%
2012	13,41%	4,87%	7,65%	8,29%	8,33%	8,03%	7,78%
2013	29,60%	3,64%	−13,94%	0,50%	−10,14%	−5,21%	2,58%
2014	11,39%	6,52%	12,74%	25,68%	20,72%	19,27%	24,30%
2015	−0,73%	2,70%	8,75%	17,01%	12,14%	11,00%	14,63%
2016	9,54%	2,18%	13,28%	7,31%	9,07%	16,05%	3,34%
2017	19,42%	3,09%	−5,01%	2,24%	−2,37%	1,93%	2,98%
2018	−6,24%	4,81%	41,10%	43,30%	38,71%	35,52%	32,36%
Rate	6,26%	4,06%	8,55%	14,15%	10,25%	11,42%	10,83%
ER	0	−2,19%	2,29%	7,90%	4,00%	5,17%	4,57%
SR	0,52	3,04	0,57	1,07	0,75	1,00	1,01
Panel C: All period 2000–2018							
Rate	3,14%	4,49%	15,06%	14,15%	14,83%	14,69%	12,90%
ER	0	1,35%	11,92%	11,02%	11,70%	11,55%	9,77%
SR	0,75	4,00	0,76	1,15	1,00	0,96	1,43
AVG	4,51%	4,50%	16,64%	14,78%	15,75%	15,64%	13,26%
MED	8,99%	4,76%	12,74%	10,67%	12,16%	16,05%	12,82%
MIN	−38,49%	2,18%	−13,94%	0,50%	−10,14%	−8,10%	2,58%
MAX	29,60%	6,52%	73,67%	43,30%	54,60%	59,50%	32,36%
(+)	12	19	16	19	16	17	19
(−)	7	0	3	0	3	2	0

Notes: SR = Sharp Ratio; ER = Excess return (comparing to the benchmark); AVG = Average annually returns; MED = Median annually returns; MIN = Minimum annually returns; MAX = Maximum annually returns; (+) count of positive years; (−) count of negative years.

Table 5.
Results from the 6 models.

	Model 1 Max SR	Model 2 Max return	Model 3 Equal return	Model 4 Max return/ SR	Model 5 Model 4 (ver2)	Model 6 Max min
SP 500						
DJ 30	6%					
NASDAQ						
STOX 600						
HANG SENG						
Emerging Mkts						
RE		18%	11%	31%	16%	19%
Consumer	9%		23%		3%	3%
Healthcare	1%					
Communications						
Energy						
Financials						
Industrials						
Semiconductors			9%	2%	16%	16%
B 1-3y	32%					
B 7-10y						1%
B 20 + y						1%
TIPS						2%
Corporate B						
BUND	33%					1%
Cocoa	3%		2%	2%		17%
Coffee						
Corn						
Sugar	1%					
GOLD						3%
Cooper						
Silver	2%	35%	19%	26%	27%	11%
Crude						
Natural Gas		16%	4%	9%	11%	3%
Commodities						
EUR/USD	9%					
VIX	3%	31%	32%	29%	28%	21%

Notes: SR = Sharp Ratio; SPX = Standard & Poor's ticker; DJI = Dow Jones Industrials ticker; STOXX = Eurostoxx 600 ticker; HSI = Hang Seng Index; EM = Emerging Markets; RE = Real Estate; B-ST = US Bonds Short Term; B-MT = US Bonds Medium Term; B-LT = US Bonds Long Term; TIP = Treasury Inflation Protected Securities; CorpB = Corporation Bonds; NGas = Natural Gas; EUR/USD = Euro vs. USA Dollars exchange; VIX = Volatility Index (S&P 500).

Table 6.
Constitution of the 6 models.

because metals came across with a big drop which is a big part of the constitution of this portfolio (38% commodities, which 26% silver). The remain constitution: 33% equity and 29% VIX. It is seen 3 negative years (2001, 2013 and 2017) which is

	S&P 500	Portfolio 1	Portfolio 2	Portfolio 3	Portfolio 4	Portfolio 5	Portfolio 6
S&P 500	1						
Portfolio1	0,11	1					
Portfolio2	−0,23	0,30	1				
Portfolio3	−0,09	0,41	0,81	1			
Portfolio4	−0,18	0,40	0,97	0,89	1		
Portfolio5	0,04	0,35	0,94	0,86	0,95	1	
Portfolio6	−0,06	0,51	0,69	0,89	0,81	0,77	1

Table 7.
Correlation matrix of portfolios vs. benchmark.

perfectly acceptable when comparing to the benchmark (7 negative years in 19 years total). Portfolio 5 is very similar to Portfolio 4, in results and in constitution. Here, the difference is to assure a sharp ratio equal or superior to 1 for the first decade and for the second decade as well. There was a little improvement comparing to the last one, the model will "steal" 1% of the returns from the first decade and return it to the second decade, i.e., instead of 18,28% vs. 10,25% (Portfolio 4), we get 17,12% vs. 11,42%. Also, instead of 3 negative years, there is 2 negative years (2001 and 2013) and the worst year instead of −10,14% (Portfolio 4), −8,10%. Finally, Portfolio 6 we maximize the minimum return (yearly). We may say that this portfolio is an upgrade from the first one (Portfolio 1), because 1. there are no negative years, 2. the worse year presents a positive return of 2,58% and 3. it maximizes more returns to the investor. The overall return is 12,9% yearly (vs 4,49% - Portfolio 1) and sharp ratio is superior to 1 for both decades. What regards to its constitution: 38% equities, 35% commodities, 21% VIX and 5% treasury bonds.

As can be seen, all portfolios come across to the benchmark, portfolio 1 with less spikes, although, S&P500 is almost touching the line of the portfolio. Portfolio 2 seems to be the most volatile. **Table** 7 shows the correlation matrix of portfolios and benchmark between themselves.

As can be seen, there is no correlation between S&P 500 and any portfolio, meaning that our proposed portfolios behave quite independently from the stock market. Portfolio 1, where we maximize the sharp ratio has no correlation with others 5 portfolios at all. Portfolio 2 to 5 are highly correlated between themselves and Portfolio 6 (max min) is highly correlated to Portfolios 3 to 5.

6. Conclusions

Our study shows that is possible to create robust portfolios where the risk is minimized, and the return is maximized. Theory behind is [2] which study focus on 'efficient frontier of optimal investment', while advocating a diversified portfolio to reduce risk. To perform it, six portfolio models are proposed, and its formation are made by a solver, where the selected solving method is the GRG Nonlinear engine for linear solver problems. Then we compare results with the benchmark (S&P 500), a linear regression model (available for request) and other "popular" portfolios (already known by many investors – also, only available by request) as well.

Results show that the GRG Nonlinear engine is powerful, providing excess returns to all six models. We design models for three types of investors: conservative, moderate and aggressive. For a conservative investor, portfolio 1 fits the best followed by portfolios 6 and 3. Portfolio 1 shows a strong sharp ratio equal to 4, presenting though, very low volatility but with lower returns when comparing to portfolios 6 and 3. None of this mentioned portfolios show a negative year during the period of 2000 to 2018. Portfolio 3 presents a high performance (14,15% annually). For an aggressive investor portfolio 2 is the best choice because it maximizes the overall return. It is the most volatile portfolio, but it may generate an average income superior to 15% annually. For a moderate investor, fan is wider but still we would exclude the portfolio 1 because it will not generate to much return and portfolio 2 may be a little volatile. Even that, portfolio 2 only presents three negatives years which is still better than the benchmark (3 versus 7).

We went further in our research and we figure out that GRG is robust, and its returns exceeds the other models mentioned in the first paragraph of this section (linear and "classical" portfolios).

Our contribution for this study is to provide a wider variety of portfolios that can be easily used by institutional and private investors and considering that nowadays there are plenty ETFs or funds available in the market is easy to everyone to apply one of the proposed models. Also, it is proved that it is possible to design very efficient portfolios, increasing returns and at the same time, lowering the risk. The results enforce the MPT from [2, 3].

As it happens in all models, there are, of course, some limitations as well. First, we may not guarantee that portfolios constitutions (1–6) will present the same results in the future because we are relaying in past returns and we would need, at least one more decade to understand if, for example, "good" decades present similar behavior between themselves. Another limitation found, the lack of the real VIX tracker (ETF/ETN). Available ETFs of VIX are a mix of mid-short term that do not reflect the actual index. Note that VIX plays an important role in GRG models.

Acknowledgements

The views expressed in this paper are those of the authors and do not necessarily represent the views of the institutions with which they are affiliated. The authors acknowledge research from the FCT (NECE: UID/GES/04630/2019). The FCT NOVA's author acknowledges Fundação para a Ciência e a Tecnologia (FCT - MCTES) for its financial support via the project UIDB/00667/2020 (UNIDEMI).

Author details

Raúl D. Navas[1,2], Sónia R. Bentes[1,3] and Helena V.G. Navas[4,5*]

1 Lisbon Accounting and Business School (LABS) – Lisbon Polytechnic Institute (IPL), Lisbon, Portugal

2 Research Center in Business Sciences (NECE), Covilhã, Portugal

3 BRU - The Business Research Unit, Lisbon University Institute, Portugal

4 NOVA School of Science and Technology, Universidade NOVA de Lisboa, Caparica, Portugal

5 UNIDEMI - Research and Development Unit in Mechanical and Industrial Engineering, FCT NOVA, Caparica, Portugal

*Address all correspondence to: hvgn@fct.unl.pt

References

[1] Robbins T. Money: Master the Game. New York: SIMON & SCHUSTER LTD.; 2014

[2] Markowitz H. Portfolio Selection: Efficient Diversification of Investments. New York: Wiley; 1959

[3] Markowitz, H. (2010). Portfolio theory: As I still see it. Annual Review of Financial Economics.

[4] Hirshleifer D. Behavioral Finance. Annual Review of Financial Economics. 2015;7:133-159

[5] Lusardi A, Mitchell O. Financial literacy around the world: an overview. Journal of Pension Economics & Finance. 2011;**10**:497-508

[6] Graham B, Zweig J, Buffett WE. The Intelligent Investor: The Definitive Book on Value Investing. London: Paperback; 2003

[7] Gorton G. Financial Crises. Annual Review of Financial Economics. 2018;**10**:43-58

[8] Timmermann A. Forecasting Methods in Finance. Annual Review of Financial Economics. 2018;**10**:449-479

[9] Daniel K, Hirshleifer D, Subrahmanyam A. Investor Psychology and Security Market Under- and Overreactions. Journal of Finance. 1998;**53**:1839-1885

[10] Debont, W., & Thaler, R. (1995). Financial Decision-Making in Markets and Firms: A Behavioral Perspective. In V. M. R.A. Jarrow, Handbooks in Operations Research & Management Science, Vol. 9: Finance.

[11] Hirshleifer D. Do individual investors drive post-earnings announcement drift? Accounting Review, pp. 2008:1521-1550

[12] Fama E. Market efficiency, long-term returns. and behavioural finance. Journal of Financial Economics. 1998;**49**(2):283-306

[13] Li, X., Bao, J., & Hendler, J. (2011). Fundamental Analysis Powered by Semantic Web. IEEE Symposium on Computational Intelligence for Financial Engineering and Economics, pp. 1-8.

[14] Xue Y, Zhang M. Fundamental Analysis, Institutional Investment, and Limits to Arbitrage. Journal of Business Finance & Accounting. 2011:1156-1183

[15] Bernard V, Thomas J. Evidence that stock prices do not fully reflect the implications of current earnings for future earnings. Journal of Accounting and Economics. 1990;**13**:305-340

[16] Dharmesh V, Nakul S. Contemporary issues in accounting. Post earnings announcement drift., ed 5th. 1995:269

[17] Richardson S, Tuna I, Wysocki P. Accounting anomalies and fundamental analysis: A review of recent research advances. Journal of Accounting & Economics. 2010;**50**(2-3):410-454

[18] Fanta A, Makina D. Equity, Bonds, Institutional Debt and Economic Growth: Evidence from South Africa. South African Journal of Economics. 2017;**85**(1):86-97

[19] McMillan D. The Behaviour of the Equity Yield and Its Relation with the Bond Yield: The Role of Inflation. International Journal of Financial Studies. 2018;**6**:99-117

[20] Cowles A, Jones H. Some a posteriori probabilities in stock market action. Econometrica, Volume 5. Issue. 2002;**3**(Jul., 1937):280-294

[21] Kendall M, Hill A. (2002). The analysis of economic time-series-part I. Prices. Journal of the Royal Statistical Society, Series A. 1953;**116**(1):11-34

[22] Fama EF. Efficient capital markets: a review of theory and empirical work. Journal of Finance. 1970;**25**(2):383-417

[23] Kirman A. Whom or What Does the Representative Individual Represent? Journal of Economic Perspectives. 1992;**6**(2):117-136

[24] Working H. A random-difference series for use in the analysis of time series. Journal of the American Statistical Association, Volume 29. Issue. 2002;**185**(Mar., 1934):11-24

[25] Cheng I-H, Xiong W. Financialization of Commodity Markets. Annual Review of Financial Economics. 2014;**6**:419-441

[26] Bhalla VK. Investment Management. New Delhi: S. Chand & Co. Ltd.; 2010

[27] Elton EJ, Gruber MJ. Modern portfolio theory, 1950 to date. Journal of Banking & Finance. 1997;**21**:1743-1759

[28] Merton R. An analytic derivation of the efficient portfolio frontier. Journal of Financial and Quantitative Analysis. 1972, September;**7**:1851-1872

[29] Levy H, De Giorgi E, Hens T. Two Paradigms and Nobel Prizes in Economics: a Contradiction or Coexistence? European Financial Management. 2012;**18**(2):163-182

[30] Rutterford J, Sotiropoulos D. Financial diversification before modern portfolio theory: UK financial advice documents in the late nineteenth and the beginning of the twentieth century. European Journal History of Economic Thought. 2016;**23**(6):919-945

Chapter 9

Influence of Injection Molding Parameters on the Mechanical Properties of Injected PC/ABS Parts

Fatma Hentati and Neila Masmoudi

Abstract

This study optimized the influence of process parameters on the mechanical properties during injection molding (IM) of PC/ABS blend. The Taguchi method of design of experiments (DOE) was employed to optimize the process parameters and to increase the tensile strength and the elasticity module. Taguchi's L9 (3^4) orthogonal array design was employed for the experimental plan. Process parameters of the injection molding such as material temperature, injection pressure, holding time, and mold temperature were studied with three levels. The Signal to noise (S/N) ratio for mechanical properties of PC/ABS blend using the Taguchi method was calculated. Taguchi's results proposed two sets of optimal injection parameters conditions to achieve the best mechanical characteristics (σ, E). The (S/N) ratio results proved that the injection pressure was the more prominent than the other IM process parameters for the tensile strength, and the material temperature was the more prominent for the elasticity module.

Keywords: PC/ABS parts, injection molding parameters, Taguchi method, mechanical properties

1. Introduction

Plastic injection molding (IM) is one of the most important polymer processing operations in the plastic industry nowadays. The application of this process is increasing significantly in many fields, especially in the automotive parts. The most process injection molding parameters that analyze the mechanical properties are pressure, temperature, and time.

Experimental and-error approach to control the process parameters for injection molding is no longer effective. Wrong selection of these parameters induces a drop in the mechanical properties of injected molded parts [1–6].

Such problems can be solved by first developing optimization models correlating the responses and the process parameters. Optimizing these parameters requires a careful endeavor. Besides, the method of varying one parameter while keeping the others constant contains limitations such as the long time experience and the large number of experiments which raise its cost in order to get an important quality of the parts. Hence, a suitable optimization technique using design of experiment DOE by Taguchi approach is then applied to search for fine tuning of parameter values to obtain the desired responses. The Taguchi method introduces an integrated approach that is simple and efficient for finding the best range of designs for quality, performance, and computational cost [7–9]. Taguchi method is a useful methodology for analyzing the variation using the signal-to noise (S/N) ratio. The Taguchi analysis of the S/N ratio specifies three situations of quality characteristics, as well as the-larger-the-better (LB), the nominal-the-better (NB) and the-smaller-the-better (SB).

A great deal of research used this approach is being carried out to understand, identify critical factors and to optimize the molding process. For example, Wegrzyn et al. conducted a study to investigate the impact of injection speed and melt temperature on electrical conductivity and dynamic mechanical properties of PC/ABS-MWCNT Nano composites [10]. These authors figured out that the injected molded specimens recover conductivity. In addition, Wang et al. elaborated experimentally the compression behavior of PC/ABS during monotonic and cyclic loading over a wide-ranging of temperatures (up to 373 K) and strain rates (up to 5000 s^{-1}) [11]. In parallel, Li et al. analyzed effect of process parameters namely, melt temperatures, injection speed and injection pressure to determine the influence of weld lines on appearance of PP products using Taguchi experimental design method [12].

Rafizadeh et al. showed that when fixing the mixing time and temperature, temperature and pressure of injection, mold temperature, and blend composition in mixing through the Taguchi orthogonal array, the impact strength of PC/ABS blend could be optimized [13]. Equally, Ozcelik studied the influence of the injection parameters and weld line on the mechanical properties of polypropylene (PP) during IM process via the design of experiment (DOE). Mechanical properties such as maximum tensile load, extension at break and charpy impact strength (notched) of the specimens were measured [14]. Furthermore, Kuram et al. discussed the tensile and the flexural strength during injection molding of virgin and recycled PBT/PC/ABS blends using Taguchi experimental design. The results showed the impact of injection pressure, holding pressure, and injection temperature on the mechanical properties of the molded parts [15]. Aditya M. Darekar used the Taguchi approach and found that quality and mechanical properties parts are mainly caused due to improper selection of the processing conditions during production [16].

In this study, the effect of the injection molding process on the mechanical properties of the injected PC/ABS parts was investigated. The Taguchi method was used to find out the optimal combinations of the injection molding (IM) parameters that provide better mechanical properties. These results are considered interesting especially in the automotive industries. We intend to reduce the rejection rate caused by the surface quality of some injected PC/ABS parts.

2. Material and methods

2.1 Industrial context

SKG Company recognized by the injection and metallization of the PC/ABS automotive parts, discovered the occurrence of some defects in the surfaces of the injected parts. These defects caused a high rejection rate [17].

2.2 Material elaboration

The commercial PC/ABS blend used in this investigation was the T45 PG. This blend was proximately oven dried at 120°C for 4 h [17]. Its mechanical and thermal properties are exhibited in **Table 1**.

Properties	Method	PC/ABS
MVR (260°C/5 kg)	ISO 1133	12 cm^3/10 min
Melt viscosity(1000 s^{-1}/260°C)	ISO 11443	200 Pa.s
Vicat softening temperature	ISO 306	112°C
Tensile strength (50 mm/min)	ISO 527	49 MPa
Izod notched impact strength (23°C)°	ISO 180	40 kJ/m^2

Table 1.
Characteristics data of PC/ABS blend [13].

2.3 Injection machine and molded part description

As illustrated in **Figure 1(a)**, a horizontal injection molding machine "FCSHN 125SV" was used to elaborate the PC/ABS parts. Their maximum injection pressure and clamping force were 217 MPa and 125 kN, respectively. Its screw's diameter was 34 mm.

The injection mold forms three imprints presented in **Figure 1(b)**. Imprint A was the tensile specimen. Imprint B was the dynamic tensile specimen. Imprint C was the torsion specimen [17].

This study is an alternate to investigate the impact of the injection process on the mechanical properties of injected PC/ABS parts.

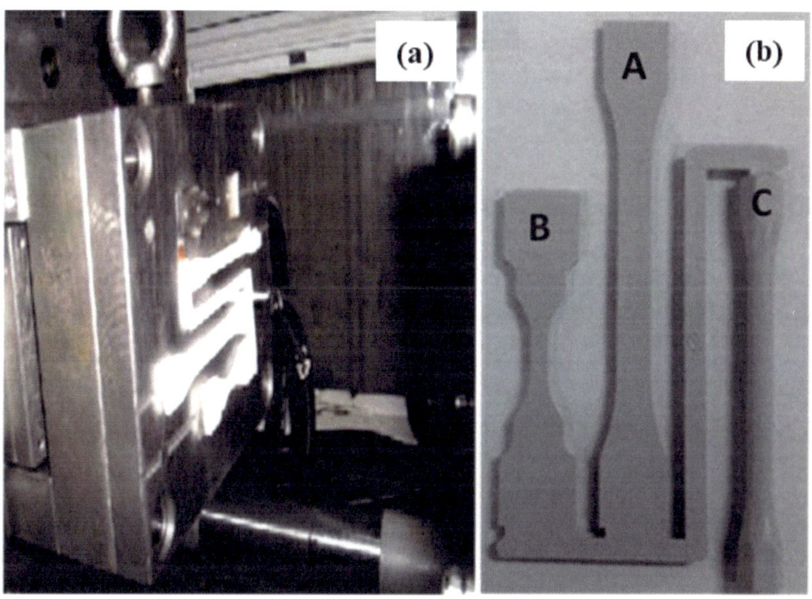

Figure 1.
(a) Injection molding machine, (b) obtained specimens [17].

2.4 Design of Experiment (DOE): Taguchi method

The developed Taguchi method goal is to determine the set of parameters which leads to the best mechanical properties of the PC/ABS injected part.

MINITAB 18 statistical software was used to find out the optimal processing conditions. A confirmation test was performed in order to guarantee the validity of the optimal experimental conditions considered from the design of the parameters.

Following preliminary tests carried out within the SKG Company, the choice of the factors and their levels was made as illustrated in **Table 2**.

The impact of the injection molding parameters on the mechanical characteristics are evaluated through the S/N ratio obtained in the Taguchi experimental plan. The S/N ratio was calculated by the larger is better quality characteristic for determining the effect of each process IM factor. The responses required are the tensile strength and the elasticity module.

Index	Factors	Unit	Level 1	Level 2	Level 3
A	Material Temperature	°C	240	250	260
B	Injection Pressure	bar	30	40	50
C	Holding Time	sec	4	8	14
D	Temperature Mold	°C	40	50	60

Table 2.
Geometric and process factors in three levels.

2.5 Tensile tests

The tensile tests were carried out, with imposed displacement, on a machine of the LLYOD INSTRUMENT type equipped with a 5 kN load cell.

All tests were achieved at ambient temperature and for a strain rate of 5 mm/min. The local deformation of the specimens during these tests was measured using an extensometer having an initial length of 25 mm and an elongation of ±1 mm.

The measurement was repeated at least 4 times for the samples and the average of three sample readings was taken for accurate results. The specimens used for the simple tensile test are taken from the ISO 527-2 standard. Its geometry is illustrated in **Figure 2**.

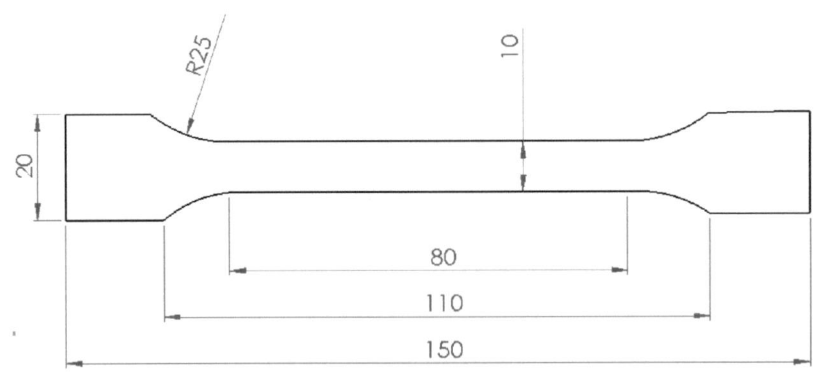

Figure 2.
Tensile specimen geometry (in mm).

2.6 Experimental

The injection molding (IM) process parameters considered in this study were: material temperature (A), injection pressure (B), holding time (C) and mold temperature (D).

In order to optimize these parameters, the famous optimization approach Taguchi was adopted. One main reason for using Taguchi is the low number of experiments. The number of experiments required could be reduced to nine, instead of 81 (3^4) experiments to find out the parameters that influence on the mechanical properties.

As depicted in **Table 3**, an orthogonal array (OA) L**9** was involved to investigate the effect of the process IM parameters. Three levels for each factor were studied. The DOF for the three levels was 2 (DOF = number of levels−1).

The elaboration of the injected PC/ABS parts was done within the SKG factory. Each combination was iterated 5 times (9 × 5 = 45). The tensile strength and the elasticity module of the injected PC/ABS parts were regarded as the desired responses.

Test Number	Material Temperature (°C)	Injection Pressure (bar)	Holding Time (sec)	Mold Temperature (°C)
1	1	1	1	1
2	1	2	2	2
3	1	3	3	3
4	2	1	2	3
5	2	2	3	1
6	2	3	1	2
7	3	1	3	2
8	3	2	1	3
9	3	3	2	1

Table 3.
Experimental layout using an L_9 orthogonal array.

3. Results and discussion

3.1 Mechanical test results

Figure 3 shows only the standard deviations of the tensile strength and elasticity module as a function of each experiment number. The mechanical

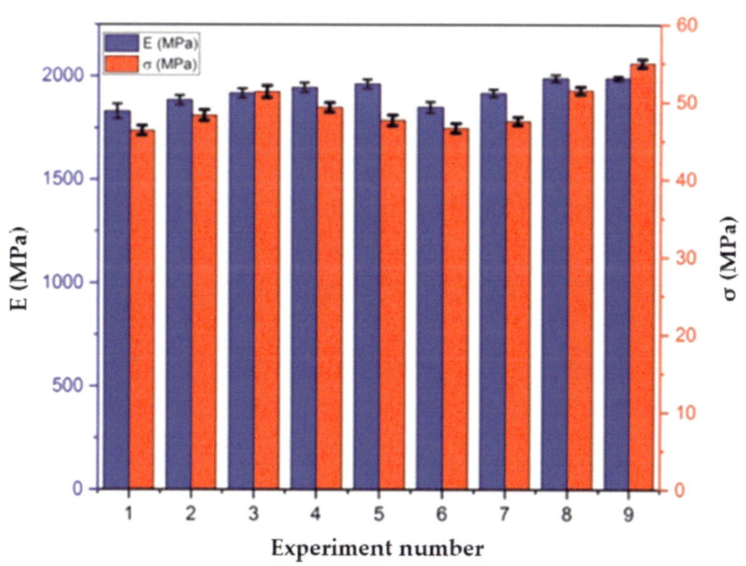

Figure 3.
Variation of the tensile strength and the elasticity module.

properties measured of each test performed were extracted. These values were introduced in the MINITAB 18 statistical software in order to optimize the injection parameters.

3.2 Taguchi results

3.2.1 Analysis of mean

The S/N ratios were calculated for each of the nine conditions of the test. All the details: the values, the average of each parameter at different levels, and the standard deviation of each test were depicted in **Table 4**.

The responses studied are the tensile strength and the elasticity module. **Tables 5** and **6** represented the average values of mean response relative to the four control parameters (factors) studied. **Figures 4** and **5** illustrated these values.

The optimum conditions corresponded to the maximum tensile strength and the elasticity module. These conditions illustrated the levels of the highest mean responses values chosen for each IM process parameter.

The averages values of the tensile strength (σ) and the elasticity module (E) were depicted in **Figures 4** and 5. Hence, the optimum condition levels corresponding to the maximum tensile strength of the injection molding process of the PC/ABS blend was A3, B3, C2, and D3. That is to say, in the third level, at the material temperature of 260°C, represented by parameter A; an injection pressure of 50 bar, represented by parameter B; a mold temperature of 60°C represented by parameter D; and in

Test Number	A	B	C	D	σ(MPa)	S/N	E(MPa)	S/N
1	1	1	1	1	46.46	33.34	1830	65.24
2	1	2	2	2	48.3	33.67	1885	65.5
3	1	3	3	3	51.38	34.21	1918	65.65
4	2	1	2	3	49.33	33.86	1946	65.78
5	2	2	3	1	47.66	33.56	1962.67	65.85
6	2	3	1	2	46.66	33.37	1850	65.34
7	3	1	3	2	47.53	33.53	1917.67	65.65
8	3	2	1	3	57.5	34.23	1989.67	65.97
9	3	3	2	1	55	34.81	1990	65.97

$\bar{\sigma}$: Overall mean of tensile strength = 49.31 MPa.
$\bar{E}$: Overall mean of elasticity module = 1921 MPa.

Table 4.
Experimental results of mechanical properties for T45PG.

Parameter	Level 1	Level 2	Level 3	Max-Min	Rank
A	48.72	47.89	51.36	3.47	1
B	47.78	49.16	51	3.25	2
C	48.21	50.89	48.86	2.68	4
D	49.72	47.5	50.74	3.24	3

Table 5.
Average values of the tensile strength (σ) at the different levels and their main effects.

Parameter	Level 1	Level 2	Level 3	Max-Min	Rank
A	1878	1920	1966	88	1
B	1898	1946	1919	48	4
C	1890	1940	1933	50	3
D	1928	1884	1951	67	2

Table 6.
Average values of the elasticity module (E) at the different levels and their main effects.

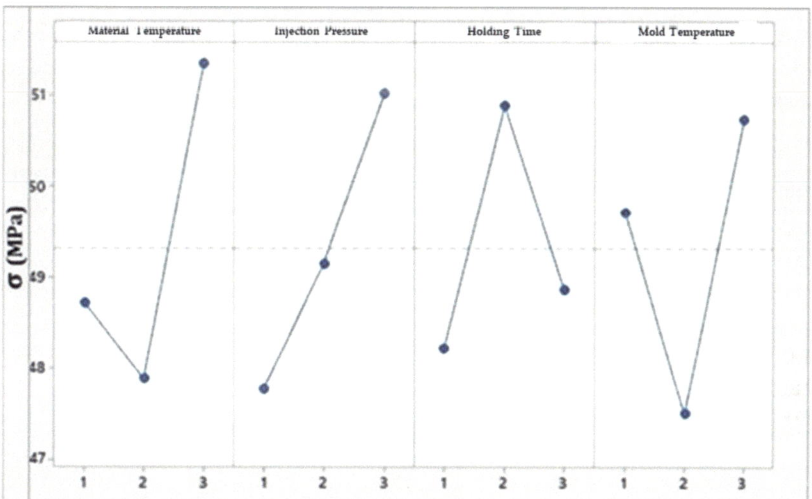

Figure 4.
Average values of tensile strength (σ) for each parameter at levels 1–3.

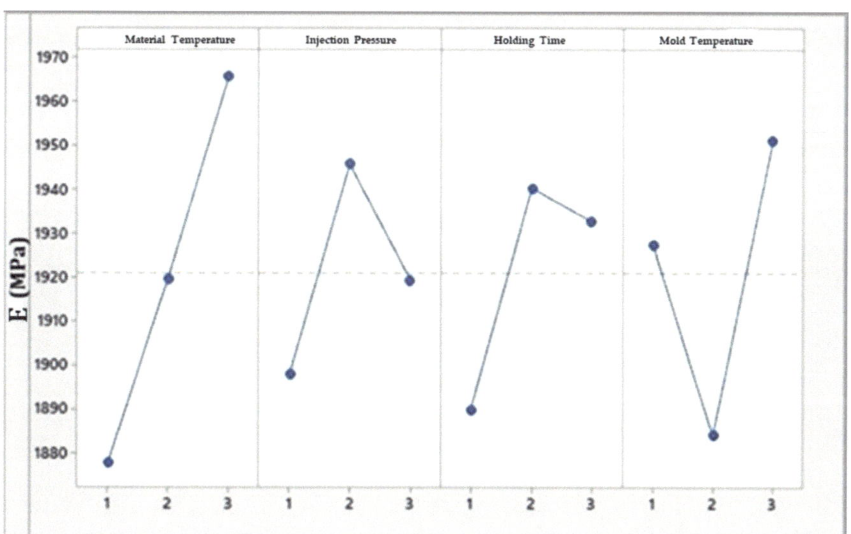

Figure 5.
Average values of elasticity module (E) for each parameter at levels 1–3.

the second level at a holding time of 8 sec, represented by parameter C. In parallel, the injection pressure was considered the more prominent than the other IM process parameters for the tensile strength.

Figure 5 shows that the material temperature represented by parameter A was more prominent than the other IM process parameters for the elasticity module. Moreover, the maximum elasticity module was recorded in the third level, at the material temperature of 260°C, represented by parameter A; a mold temperature of 60°C, represented by parameter D; and in the second level at; an injection pressure of 40 bar, represented by parameter B; a holding time of 8 sec, represented by parameter C.

The average values of S/N ratios of various parameters at different levels are shown in **Tables 7** and **8**. These values are plotted in **Figures 6** and 7.

The S/N ratio analysis of the tensile strength plotted in **Figure 6** provides in to the same levels of the parameters as depicted in **Figure 4**. Therefore, A3, B3, C2, and D3 were the best levels for reducing the variability of the injection molding process of the PC/ABS blend.

Parameter	Level 1	Level 2	Level 3	Max-Min	Rang
A	33.74	33.6	34.2	0.59	1
B	33.58	33.82	34.13	0.55	3
C	33.65	34.12	33.77	0.46	4
D	33.91	33.53	34.1	0.57	2

Table 7.
Average values (σ) of S/N ratios at the different levels and their main effects.

Parameter	Level 1	Level 2	Level 3	Max-Min	Rang
A	65.47	65.66	65.87	0.4	1
B	65.56	65.78	65.66	0.22	4
C	65.52	65.75	65.72	0.23	3
D	65.69	65.5	65.8	0.3	2

Table 8.
Average values (E) of S/N ratios at the different levels and their main effects.

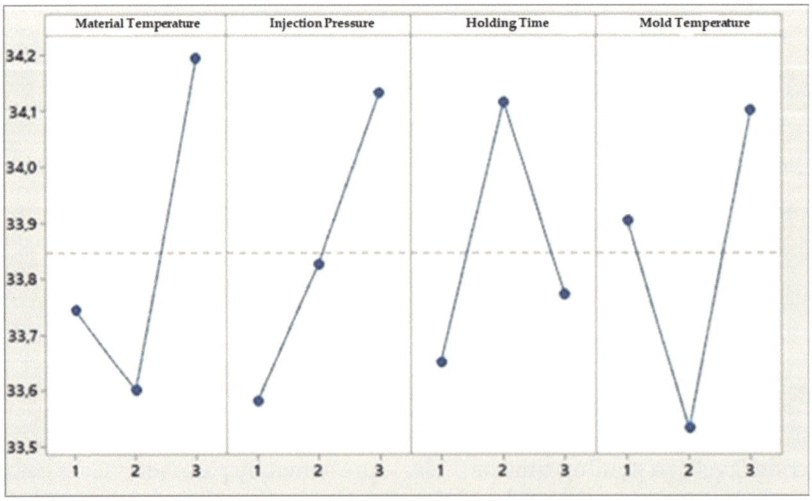

Figure 6.
Average values (σ) of S/N ratio for each parameter at levels 1–3.

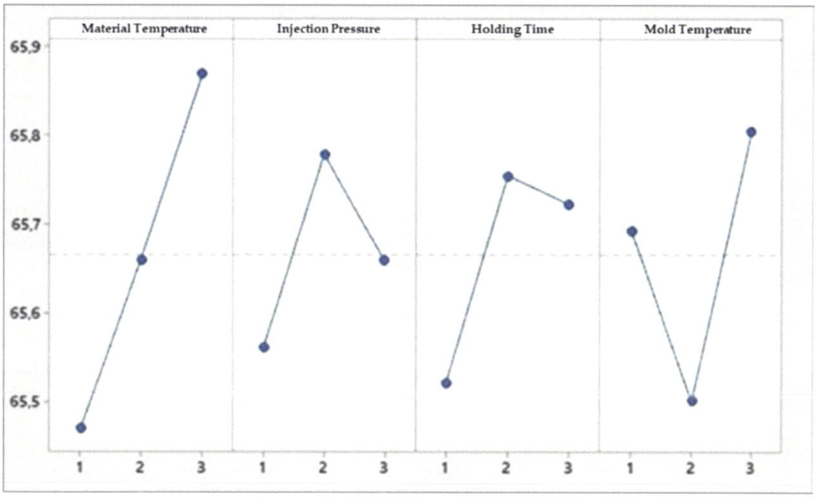

Figure 7.
Average values (E) of S/N ratio for each parameter at levels 1–3.

The S/N ratio analysis of the elasticity module depicted in **Figure 7** offers in to the same levels of the parameters as shown in **Figure 5**. So, A3, B2, C2, and D3 were the best levels for reducing the variability of the injection molding process of the PC/ABS blend.

Table 3 shows the combinations of factor levels (3, 3, 2, 3) and (3, 2, 2, 3) were not among the obtained combinations within the tested experiment. Thus, it was necessary to verify the optimum injection molding conditions via confirmation tests of the experimental design.

3.2.2 Estimation of predicted mean

The optimum levels of the control parameters of the tensile strength (μ_1) and the elasticity module (μ_2) are determined as A3, B3, C2, D3 and A3, B2, C2, D3 respectively [18, 19]:

$$\begin{aligned}\mu_1 &= \bar{\sigma} + (\bar{A}_3 - \bar{\sigma}) + (\bar{B}_3 - \bar{\sigma}) + (\bar{C}_2 - \bar{\sigma}) + (\bar{D}_3 - \bar{\sigma}) \\ &= 51.36 + 51 + 50.89 + 50.74 - (3\,\text{x}\,49.31) = 56.06\,MPa.\end{aligned} \tag{1}$$

Where $\bar{A}_3$ is the average tensile strength at the third level of material temperature, 260°C, $\bar{B}_3$ is the average tensile strength at the third level of injection pressure, 50 MPa, $\bar{C}_2$ is the average tensile strength at the third level of holding time, 8 sec, $\bar{D}_3$ is the average tensile strength at the third level of mold temperature, 60°C and $\bar{\sigma}$ is the mean of tensile strength obtained from **Table 4**.

$$\begin{aligned}\mu_2 &= \bar{E} + (\bar{A}_3 - \bar{E}) + (\bar{B}_2 - \bar{E}) + (\bar{C}_2 - \bar{E}) + (\bar{D}_3 - \bar{E}) \\ &= 1966 + 1946 + 1940 + 1951 - (3\,\text{x}\,1921) = 2040\,MPa.\end{aligned} \tag{2}$$

Where $\bar{A}_3$ is the average elasticity module at the third level of material temperature, 260°C; $\bar{B}_2$ is the average elasticity module at the second level of injection pressure, 40 MPa, $\bar{C}_2$ is the average elasticity module at the second level of holding time, 8 sec, $\bar{D}_3$ is the average elasticity module at the third level of mold temperature, 60°C and $\bar{E}$ is the mean of elasticity module obtained from **Table 4**.

3.2.3 Confirmation tests

Three confirmation experiments were conducted at the optimum settings of the process parameters recommended by the investigation. The average tensile strength and the elasticity module obtained at the optimal level of the process parameters were 56.28 MPa and 1983 MPa.

Obviously, there was a difference between the computed and the experimental results. However, this difference can be considered not significant. Therefore, the confirmation tests indicate that the optimal conditions obtained above produced the best mechanical properties (σ, E) of the PC/ABS blend (**Table 9**).

	Experimental value	Measured value	Error
Tensile Strength (σ)	56.06 MPa	56.28 MPa	0.39%
Elasticity module (E)	2040 MPa	1983 MPa	2.28%

Table 9.
Confirmation test results.

4. Conclusion

This work focused on Taguchi experimental method for investigating the influence of the injection parameters on the mechanical properties of PC/ABS during injection molding. Taguchi's results proposed two sets of optimal injection parameters conditions to achieve the best mechanical characteristics (σ, E). A first series: a material temperature of 260°C, an injection pressure of 50 bars, a holding time of 8 sec and a mold temperature of 60°C. A second series: a material temperature of 260°C, an injection pressure of 40 bar, a holding time of 8 seconds and a mold temperature of 60°C. All mechanical properties measurements matched very well with the experimental data. The most important parameter affecting the maximum tensile strength was the injection pressure. However, the material temperature was considered the most important parameter affecting the elasticity module. Consequently, it is shown clearly the above performance characteristics in the injection molding process are greatly significant through this study. So, we recommend that future works involve the impact of injection processing on the surface quality of the PC/ABS parts.

Acknowledgements

The authors are grateful for the help of the SKG staff during the materialization of this study.

Acronyms and abbreviations

IM	Injection Molding
DOE	Design of Experiment
S/N	Signal-to-Noise ratio

OA	The Orthogonal Array
T_{ma}	The material temperature
T_{mo}	The mold temperature
P_{inj}	The injection pressure
t_h	The holding time
MPa	Mega Pascal
°C	Celsius temperature scale
bar	Bar pression scale
sec	second time scale

Author details

Fatma Hentati[1]* and Neila Masmoudi[2]

1 Laboratory of Applied Mechanics and Engineering (MAI), National School of Engineers of Tunis, Tunis, Tunisia

2 Laboratory of Electromechanical Systems (LASEM), National School of Engineers of Sfax, Sfax, Tunisia

*Address all correspondence to: fatma.hentati1@gmail.com

References

[1] Jan M, Khalid MS, Awan AA, Nisar S: Optimization of injection molding process for sink marks reduction by integrating response surface design methodology & taguchi approach. J Qual Technol Manag. 2016; Volume XII, Issue I:45-79.

[2] Chen WC, Nguyen MH, Chiu WH, Chen TN, Tai PH: Optimization of the plastic injection molding process using the Taguchi method, RSM, and hybrid GA-PSO. Int J Adv Manuf Technol. 2016; 83:1873-1886. https://doi.org/10.1007/s00170-015-7683-0

[3] Tang S.H., Tan Y.J., Sapuan S.M., Sulaiman S., Ismail N, Samin R: The use of Taguchi method in the design of plastic injection mould for reducing warpage. Journal of Materials Processing Technology. 2007; 182: 418-426. https://doi.org/10.1016/j.jmatprotec.2006.08.025

[4] Ozcelik B, Sonat I: Warpage and Structural analysis of thin shell plastic in the plastic injection molding. Materials and Design 30: 367-375. https://doi.org/10.1016/j.matdes.2008.04.053

[5] Huang M.S, Lin T.Y: Simulation of a regression-model and PCA based searching method developed for setting the robust injection molding parameters of multi-quality characteristics. International Journal of Heat and Mass Transfer. 2008; 51: 5828-5837.

[6] Cardozo D: Three models of the 3D filling simulation for injection Molding: A Brief Review. Journal of Reinforced Plastics and Composites. 2008; *27*, 1963.

[7] Taguchi G: Introduction to Quality Engineering. Mc Graw-Hill, New York. 1990.

[8] Peace G.S: Taguchi Methods, Addison-Wesley Publishing Company: Taipei. 1993; 281-291.

[9] Dwiwedi AK, Kumar S, Rahbar NN, Kumar D: Practical application of Taguchi method for optimization of process parameters in injection molding machine for PP material. International Research Journal of Engineering and Technology (IRJET) e-ISSN. 2015; 2395-0056.

[10] Wegrzyn M, Juan S, Benedito A, Giménez E: The influence of injection molding parameters on electrical properties of PC/ABS-MWCNT nanocomposites. J. Appl.Polym. Sci. 2013; Volume 130: 2152-2158. https://doi.org/10.1002/app.39412

[11] Wang H, Zhou H, Huang Z, Zhang Y, Qiao H, Yu Z: Polym. Test. 2016; 50: 216.

[12] Li H, Guo Z, Li D: Reducing the effects of weld lines on appearance of plastic products by Taguchi experimental method. International Journal of Advanced Manufacturing Technology. 2007; 32:927-931.

[13] Rafizadeh M, Morshedian J, Ghasemi E, Bolouri A. Iran Poly J. 2005; 14: 881.

[14] Ozcelik B: Optimization of injection parameters for mechanical properties of specimens with weld line of polypropylene using Taguchi method. Int Commun Heat Mass Transf. 2011; 38:1067-1072. https://doi.org/10.1016/j.icheatmasstransfer.2011.04.025

[15] Kuram E, Timur G, Ozcelik B, Yilmaz F. Mater. Manuf. Proc. 2014; 29, 1260.

[16] Aditya M. Darekar, Venkatesh T. S, Bhushan T. Patil, Yazad N: Review of Optimization Aspects for Plastic Injection Molding Process. IRACST – Engineering Science and Technology: An International Journal (ESTIJ). 2015; ISSN: 2250-3498 Vol.5, No.1.

[17] Hentati F, Hadriche I, Masmoudi N, Bradai C: Optimization of the injection molding process for the PC/ABS parts by integrating Taguchi approach and CAE simulation. The International Journal of Advanced Manufacturing Technology. 2019; ISSN 0268-3768. https://doi.org/10.1007/s00170-019-04283-z

[18] Neşeli S, Yaldiz S, Türkeş E: Optimization of tool geometry parameters for turning operations based on the response surface methodology. Meas J Int Meas Confed. 2011; 44:580-587. https://doi.org/10.1016/j.measurement.2010.11.018

[19] Lakshminarayanan AK, Balasubramania V: Process parameter optimization for friction stir welding of aluminium 2014-t651 alloy using taguchi technique. Trans Nonferrous Met Soc China. 2008; 18:548-554.

Chapter 10

Application of Taguchi Method in Optimization of Pulsed TIG Welding Process Parameter

Asif Ahmad

Abstract

Pulsed TIG welding is one of the most widely used welding processes in the metal manufacturing industry. In any fusion arc welding process, the bead width plays an important role in determining the welding strength and mechanical properties of the weld joint. This study present optimization of the pulsed TIG welding process parameter using Taguchi Philosophy. AISI 316/3136L austenite stainless steel 4mm is used for welding and for the establishment of the optimum combination of the process parameter and depending upon the functional requirement of the welded joint, the acceptable welded joint should have optimum bead width and minimum heat affected zone (HAZ) etc. An experiment was conducted using different welding condition and a mathematical model was constructed using the data collected from the experiment based on Taguchi L_{25} orthogonal array. Optimum parameter obtained for bead width is peak current 180 ampere, base current 100 ampere, pulse frequency 125Hz and pulse on time 40%.

Keywords: TIG welding, design of experiment, Taguchi methodology, S/N ratio, ANOVA

1. Introduction

After the Second World War, the associated powers found that the nature of the Japanese telephone system was incredibly poor and absolutely unacceptable for long term communication purposes. To improve the system, it is recommended to establishing research facilities in order to develop a state-of-the-art communication system. The Japanese founded the Electrical Communication Laboratories (ECL) with Dr. Genichi Taguchi in charge of improving R&D efficiency and improving product quality. He observed that a great deal of time and money was expended on engineering experimentation and testing [1]. Taguchi seen quality improvement as a progressing exertion. He continually strived to reduce the variation around the target value. To accomplish this, Taguchi designed experiments using specially constructed tables known as OA. The use of these tables makes the design of experiments very easy and consistent [2]. Design of Experiments (DOE) is powerful statistical technique presented by R. A. Fisher in England during the 1920s to study the impact of numerous factors at the same time. In his initial applications, Fisher needed to discover how much rain, water, fertilizer, sunshine, etc. are expected to deliver the best yield. Since that time, much improvement of the system has occurred in the scholarly condition yet helped create numerous applications on the generation floor [3]. In late 1940s Dr. Genechi Taguchi of Electronic Control Laboratory in Japan, carried out significant research with DOE techniques. He spent extensive exertion to make this trial procedure easier to use and to improve the quality of manufactured products. Dr. Taguchi's standardized version of DOE, popularly known as the Taguchi method or Taguchi approach, was introduced in the USA in the early 1980s. Today it is one of best optimization techniques used by manufacturing industry. The DOE using the Taguchi approach can monetarily satisfy the needs of problem-solving and product/process design in optimization projects. By learning and applying this procedure, specialists, researchers, and scientists can essentially decrease the time required for exploratory examinations [4].

2. Taguchi approach

Design of Experiments (DOE) is powerful statistical technique presented by R. A. Fisher in England during the 1920s to study the impact of numerous factors at the same time. In his initial applications, Fisher needed to discover how much rain, water, fertilizer, sunshine, etc. are expected to deliver the best yield. Since that time, much improvement of the system has occurred in the scholarly condition yet helped create numerous applications on the generation floor [3]. In late 1940s Dr. Genechi Taguchi of Electronic Control Laboratory in Japan, carried out significant research with DOE techniques. He spent extensive exertion to make this trial procedure easier to use and to improve the quality of manufactured products. Dr. Taguchi's standardized version of DOE, popularly known as the Taguchi method or Taguchi approach, was introduced in the USA in the early 1980s. Today it is one of best optimization techniques used by manufacturing industry. The DOE using the Taguchi approach can monetarily satisfy the needs of problem-solving and product/process design in optimization projects. By learning and applying this procedure, specialists, researchers, and scientists can essentially decrease the time required for exploratory examinations [4].

2.1 Orthogonal array

The orthogonal array is selected as per standard orthogonal given in **Table 1**. This technique was first given by Sir R. A. Fisher, in the 1920s [5]. The method is popularly known as the factorial DOE. A full factorial design results may involve a large number of experiments. A full factorial experiment as shown in **Table 2**.

2.2 Nomenclature array

Orthogonal array is defined as: Lx (Ny)
Where, L = Latin square
x = number of rows
N = number of levels
y = number of columns (factors)
Degrees of freedom associated with the OA = x – 1
Some of the standard orthogonal arrays are listed in **Table 3**.

- Level 1 and 2 in the matrix represent the low and high level of a factor respectively.

- Each column of the matrix has an equal number of 1 and 2.

Orthogonal array	Number of rows	Maximum no. of factor	Maximum no. of columns at these levels			
			2	3	4	5
L4	4	3	3	—	—	—
L8	8	7	7	—	—	—
L9	9	4	—	4	—	—
L12	12	11	11	—	—	—
L16	16	15	15	—	—	—
L'16	16	5	—	—	5	—
L18	18	8	1	7	—	—
L25	25	6	—	—	—	6
L27	27	13	1	13	—	—
L32	32	31	31	—	—	—
L'32	32	10	1	—	9	—
L36	36	23	11	12	—	—
L'36	36	16	3	13	—	—
L50	50	12	1	—	—	11
L54	54	26	1	25	—	—
L64	64	63	63	—	—	—
L'64	64	21	—	—	21	—
L81	81	40	—	40	—	—

Table 1.
Standard orthogonal.

Experiment no.	A	B	C
1	1	1	1
2	1	1	2
3	1	2	1
4	1	2	2
5	2	1	1
6	2	1	2
7	2	2	1
8	2	2	2

Table 2.
Full factorial experiments table.

- Any pair of columns has only four combinations [1, 1], [1, 2], [2, 1], and [2, 2] indicating that the pair of columns are orthogonal.

2.3 Signal to noise ratio

Taguchi method stresses the necessity of studying the response variable using the signal-to-noise ratio, resulting to decrease the effect of quality characteristic

Trial no.	**Column**						
	1	**2**	**3**	**4**	**5**	**6**	**7**
1	1	1	1	1	1	1	1
2	1	1	1	2	2	2	2
3	1	2	2	1	1	2	2
4	1	2	2	2	2	1	1
5	2	1	2	1	2	1	2
6	2	1	2	2	1	2	1
7	2	2	1	1	2	2	1
8	2	2	1	2	1	1	2

Table 3.
Standard L_8 orthogonal array.

variation due to the uncontrollable parameter. The S/N ratio can be used in three types:

i. Larger the better:

$$S/N\ Ratio = -10\log.\ 1/a\ [\textstyle\sum_{i=0} 1/yi2]$$

ii. Smaller the better:

$$S/N\ Ratio = -10\log.1/a\ [\textstyle\sum_{i=0} yi2]$$

iii. Nominal the best:

$$S/N\ Ratio = -10\log.\ [\textstyle\sum_{i=\bar{0}} yi\ 2/s2]$$

Where,
a = Number of trials
yi = measured value
ȳ = mean of the measured value
s = standard deviation
Parameters that affect the output can be divided into two parts: controllable (or design) factors and uncontrollable (or noise) factors. Uncontrollable factors cannot be controlled but its effect can be minimized by varying the controllable factors.

2.4 Analysis of variance

ANOVA were first introduced by Sir Ronald A, Fisher, the British biologist. ANOVA is a method of partitioning total variation into accountable sources of variation in an experiment. It is a statistical method used to interpret experimented data and make decisions about the parameters under study. ANOVA is a statistical method used to test differences between two or more means [1].

2.4.1 Hypotheses of ANOVA

H0: The (population) means of all groups under consideration are equal.
Ha: The (pop.) means are not all equal. (Note: This is different than saying. they are all unequal.)

2.4.2 ANOVA table

A detail of all analysis of variance computations is given **Table 4**.
Where,
N = total number of observations
SSf = sum of squares of a factor
K = number of levels of the factor
SSe = sum of squares of error
Fo = computed value of F
Vf = variance of the factor
Ve = variance of the error

2.4.3 One-way ANOVA and their notation

When there is just one explanatory variable, we refer to the analysis of variance as a one-way ANOVA.

Here is a key to symbols you may see as you read through this section.
k = the number of groups/populations/
xi j = the jth response sampled from the ith group/population.
$\overline{\text{xi}}$ = the sample mean of responses from the ith group $= \frac{1}{\text{ni}}\sum_{\text{j=i}}^{\text{n}}\text{xij}$
si = the sample standard deviation from the ith group $= 1/(\text{ni}-1)\sum_{\text{j=1}}^{\text{ni}}(\overline{\text{x}}\text{ij-xi})^2$
n = the total sample $= \sum_{\text{i=0}}^{\text{k}}\text{xi}$
$\overline{\text{x}}$ = the mean of all responses $= 1/\text{n}\sum_{\text{ij}}\text{xij}$

2.4.4 Parting the total variability

Viewed as one sample one might measure the total amount of variability among observations by summing the squares of the differences between each xi j and x:

Sources of variability:

1. SST (stands for the sum of squares total) $\sum_{\text{j=1}}^{\text{ni}}\cdot\sum_{\text{j=1}}^{\text{ni}}(\text{xij}-\overline{\text{x}})^2$

2. Sum of Square Group between group

   SSG = $\sum_{\text{i=0}}^{\text{k}}\text{ni}(\text{xij}-\text{x})^2$

   Sum of Square Group within groups means

3. SSE = $\sum_{\text{j=1}}^{\text{ni}}\cdot\sum_{\text{j=1}}^{\text{ni}}(\text{xij}-\text{x})2 = \sum_{\text{i=1}}^{\text{k}}(\text{ni}-1)\text{s}_{\text{i}}^2$

   It is the case that SST = SSG + SSE.

Source of variation	Sum of squares	Degree of freedom	Mean square variance	Fo
Factor	SSf	K − 1	Vf = SSf/K − 1	Vf/Ve
Error	SSe	N − K	Ve = SSe/N − K	
Total	SStotal	N − 1		

Table 4.
Analysis of Variance Computations (ANOVA).

2.4.5 Calculation

An F statistic is obtained from ANOVA test or a regression analysis to find out if the means between two populations are significantly different [1]. F statistics is used to decide the acceptance or rejection of null hypothesis. F value is calculated from the data, if calculated is larger than F statistics the null hypothesis is rejected. The ANOVA table showing F value is given in **Table 5**.

SS = Sum of Squares (sum of squared deviations):
SST measures the variation of the data around the overall mean $\overline{x}$
SSG measures the variation of the group means around the overall mean x
SSE measures the variation of each observation around its group mean $\overline{x}i$

- Degrees of freedom

  k − 1 for SSG

  n − k for SSE, since it measures the variation of the n observations about k group means. n − 1 for SST, since it measures the variation of all n observations about the overall mean.

- MS = Mean Square = SS/df :

- This is like a standard deviation. Its numerator was a sum of squared deviations (just like our SS formulas), and it was divided by the appropriate number of degrees of freedom.

  It is interesting to note that another formula for MSE is

  $$MSE = \frac{(n_1-1)+(n_2-1)+(n_3-1)+\dots\dots(n_k-1)s_k^2}{(n_1-1)+(n_2-1)+\dots\dots+(n_k-1)}$$

- The F statistic = MSG/MSE

If the null hypothesis is true, the F statistic has an F distribution with k-1 and n-k degrees of freedom in the numerator/denominator respectively. If the alternative hypothesis is true, then F tends to be large. We reject Ho in favor of Ha if the F

Source	SS	df	MS	F
Model/group	SSG	k − 1	MSG $\frac{SSG}{K-1}$	$\frac{MSG}{MSE}$
Residual/Error	SSE	n − k	MSG $\frac{SSE}{n-1}$	
Total	SST	n-1		

Table 5.
ANOVA table.

	df	SS	MS	F	p-value
A	I − 1	SSA	MSA	MSA/MSE	
B	J − 1	SSB	MSB	MSB/MSE	
AXB	(I − 1) (J − 1)	SSAB	MSAB	MSAB/MSE	
Error	n − IJ	SSE	MSE		
Total	n − 1	SST			

Table 6.
Two-way ANOVA table.

statistic is sufficiently large. As with other hypothesis tests, we determine whether the F statistic is large by finding a corresponding P-value.

2.4.6 Two-way ANOVA

In the two-way ANOVA model, there are two factors, each with several levels as shown in **Table 6**.

3. Taguchi design of experiment (DOE)

Taguchi DOE is a well-known factual strategy that gives a legitimate and productive technique for process optimization. The Taguchi technique enables us to improve the consistency of production. Taguchi design recognizes that not all factors that cause variability can be controlled. These uncontrollable factors are called noise factor. Taguchi design tries to identify the controllable factor that minimizes the effect of noise factors. During experimentation, you manipulate the control factor to evaluate variability that occurs and then determines the optimal control factor setting, which minimizes the process variability. A process designed with this goal produces more consistent output and performance regardless of the environment in which it is used. It is world widely used for product design and process optimization. As a result, time is reduced considerably. Taguchi DOE methodology uses an orthogonal array that gives different combinations of parameters and their levels for each experiment [6].

3.1 The layout of the experiment

The following sequence is followed while forming the experiment.

- Base and filler material selection.
- Selection of process parameters.
- Calculating the upper and lower limits process parameters.
- Selection of standard orthogonal array.
- Experiment conducted.
- Calculating optimum condition [6].

4. Selecting base material and their mechanical properties

AISI 316 stainless steel sheets of dimension 100 × 75 × 4 mm are welded autogenously with the butt joint without edge preparation [7]. The chemical

Grade 316	C	Mn	Si	P	S	Cr	Mo	Ni	N
Min.	—	—	—	—	—	16.0	2.0	10.0	
Max.	0.08	2.0	0.75	0.045	0.030	18.0	3.0	14.0	0.10

Table 7.
Chemical composition of the base material (wt %).

Tensile strength	Tensile strength (MPa) min	Yield strength 0.2% proof (MPa)	Elongation (% in 50 mm) min	Hardness	
				Rockwell HR B max	Brinell HB max
564 MPA	515	205	40	95	217

Table 8.
Mechanical properties of AISI 316 stainless steel.

Process parameter	Code	Level 1	Level 2	Level 3	Level 4	Level 5
Peak current	P	140	150	160	170	180
Base current	B	60	70	80	90	100
Pulse frequency	F	50	75	100	125	150
Pulse on time	T	35	40	45	50	55

Table 9.
Process parameters working range.

composition and mechanical properties of 316 stainless steel sheet are given in **Tables 7** and **8**. The process parameter working range is given in **Table 9**.

4.1 An orthogonal array is selected

The input process parameters selected are four, and each parameter is divided into five levels [6]. Different Standard orthogonal array used for optimization is shown in **Table 10**.

These above standard orthogonal arrays provide full information for all possible combination of input parameter. In this experimental work, four factors with their five levels are used for which the corresponding orthogonal array is L_{25} as shown in **Table 11**. Minitab 18 statistical software is used to a developed orthogonal array, response table, main effect plot for mean and S/N ratio. AVOVA is developed by Minitab 18 software to determine the % contribution of each input parameter [8].

4.2 Conduction of experiment

By putting the values of four parameters in L_{25} Orthogonal array as shown in **Table 12** [8].

4.3 Signal to noise ratio

The S/N ratio help in measuring the sensitivity of quality characteristic to external noise factor which is not under control. The highest value of S/N ratio represent more impact of the process parameter on the output performance. On the basis of characteristic three S/N ratios are available namely lower the better, higher the better and nominal the better as shown in **Table 13**. In this paper, higher the better is used for maximizing depth of penetration as shown in Eq. (1) [6].

$$\text{S/N Ratio} = -10\log 1/\text{n}\left[\sum\nolimits_{\text{i}=0} 1/\text{yi2}\right] \tag{1}$$

Orthogonal array	Number of rows	Maximum no. of factor	Maximum no. of columns at these levels			
			2	3	4	5
L4	4	3	3	—	—	—
L8	8	7	7	—	—	—
L9	9	4	—	4	—	—
L12	12	11	11	—	—	—
L16	16	15	15	—	—	—
L'16	16	5	—	—	5	—
L18	18	8	1	7	—	—
L25	25	6	—	—	—	6
L27	27	13	1	13	—	—
L32	32	31	31	—	—	—
L'32	32	10	1	—	9	—
L36	36	23	11	12	—	—
L'36	36	16	3	13	—	—
L50	50	12	1	—	—	11
L54	54	26	1	25	—	—
L64	64	63	63	—	—	—
L'64	64	21	—	—	21	—
L81	81	40	—	40	—	—

Table 10.
Standard orthogonal array.

4.4 Experiment conducted for all input parameter

Specimen 316 austenitic stainless steel is welded as per the combination of parameters given in orthogonal array L_{25}, five trails are performed for each combination of parameters for BW then average value is taken as shown in **Table 14**. S/N ratio is obtained by using Minitab 18 statistical software as shown in **Table 15**.

4.5 Response table for bead width

The response table is obtained for the S/N ratio and mean for bead width as shown in **Tables 16** and **17**. The response table is obtained by Minitab 18 statistical software which represents the significance of each individual input parameter. Delta value is obtained for peak current, base current, pulse frequency and pulse on time which is the difference between the highest value to the lowest value. The rank of the input parameter is decided as per the highest value of delta [8].

4.6 Main effect plot for bead width

The main effect plot will help to determine the optimum value of the input parameter. Main effect plot is obtained for S/N ratio and mean for bead width by using Minitab 18 statistical software [8]. The main effect plot will represent significant the level of each input parameter as shown in **Figures 1** and **2**. Optimum value to obtained optimum bead width with their significant level is given in **Table 18**.

Experiment no.	Process parameter			
	P	B	F	T
1	1	1	1	1
2	1	2	2	2
3	1	3	3	3
4	1	4	4	4
5	1	5	5	5
6	2	1	2	3
7	2	2	3	4
8	2	3	4	5
9	2	4	5	1
10	2	5	1	2
11	3	1	3	5
12	3	2	4	1
13	3	3	5	2
14	3	4	1	3
15	3	5	2	4
16	4	1	4	2
17	4	2	5	3
18	4	3	1	4
19	4	4	2	5
20	4	5	3	1
21	5	1	5	4
22	5	2	1	5
23	5	3	2	1
24	5	4	3	2
25	5	5	4	3

Table 11.
Orthogonal Array L25 (Minitab18).

4.7 Confirmatory test for bead width

After evaluating the optimal parameter settings, the next step is to predict and verify the quality performance characteristics using the optimal parametric combination. The predicted value of the bead width is estimated by using the Eq. (2). Five experiments are conducted at the optimum parameter. The result of predicted value and experimental value of bead width is shown in **Table 19**, and it represents that predicted value and experimental value are close to each other [9].

$$\eta = n_m + \sum\nolimits_{i=0}^{0} n_{im} - n_m \tag{2}$$

where,
η – predicted value
n_m - is the total mean

Experiment no.	Process parameter			
	P	B	F	T
1	140	60	50	35
2	140	70	75	40
3	140	80	100	45
4	140	90	125	50
5	140	100	150	55
6	150	60	75	45
7	150	70	100	50
8	150	80	125	55
9	150	90	150	35
10	150	100	50	40
11	160	60	100	55
12	160	70	125	35
13	140	60	50	35
14	160	80	150	40
15	160	90	50	45
16	160	100	75	50
17	170	60	125	40
18	170	70	150	45
19	170	80	50	50
20	170	90	75	55
21	170	100	100	35
22	180	60	150	50
23	180	70	50	55
24	180	80	75	35
25	180	90	100	40

Table 12.
Orthogonal array actual value.

Signal-to-noise ratio	The goal of the experiment	Data characteristics	Signal-to-noise ratio formulas
Larger is better	Maximize the response	Positive	$-10\log\frac{1}{n}$ $[\sum_{i=0}1/yi2]$
Nominal is best	Target the response and you want to base the signal-to-noise ratio on standard deviations only	Positive, zero, or negative	$-10\log.[\sum_{i=0}\bar{y}i$ $2/s2]$
Smaller is better	Minimize the response	Non-negative with a target value of zero	$-10\log.1/n$ $[\sum_{i=0}yi2]$

Table 13.
S/N ratio.

Sr. No.	BW 1 (mm) trial 1	BW 2 (mm) trial 2	BW 3 (mm) trial 3	BW 4 (mm) trial 4	BW 5 (mm) trial 5	Average BW (mm)
1	2.78	2.8	2.77	2.785	2.806	2.79
2	2.405	2.425	2.395	2.41	2.431	2.41
3	2.13	2.15	2.12	2.135	2.156	2.14
4	2.935	2.955	2.925	2.94	2.961	2.94
5	2.515	2.535	2.505	2.52	2.541	2.52
6	2.28	2.3	2.27	2.285	2.306	2.29
7	3.095	3.115	3.085	3.1	3.121	3.10
8	2.565	2.585	2.555	2.57	2.591	2.57
9	2.325	2.345	2.315	2.33	2.351	2.33
10	3.11	3.13	3.1	3.115	3.136	3.12
11	2.615	2.635	2.605	2.62	2.641	2.62
12	2.345	2.365	2.335	2.35	2.371	2.35
13	3.345	3.365	3.335	3.35	3.371	3.35
14	2.68	2.7	2.67	2.685	2.706	2.69
15	2.475	2.495	2.465	2.48	2.501	2.48
16	3.585	3.605	3.575	3.59	3.611	3.59
17	2.775	2.795	2.765	2.78	2.801	2.78
18	2.63	2.65	2.62	2.635	2.656	2.64
19	3.375	3.395	3.365	3.38	3.401	3.38
20	2.74	2.76	2.73	2.745	2.766	2.75
21	2.425	2.445	2.415	2.43	2.451	2.43
22	3.645	3.665	3.635	3.65	3.671	3.65
23	3.13	3.15	3.12	3.135	3.156	3.14
24	2.65	2.67	2.64	2.655	2.676	2.66
25	3.87	3.89	3.86	3.875	3.896	3.88

Table 14.
Experiment value bead width.

n_{im}- is the mean value ratio at the optimal level
Average bead width = 2.83 mm
$n_{bead\ width} = 2.83 + (3.152 - 2.83) + (2.950 - 2.83) + (3.068 - 2.83) + (3.027 - 2.83)$
$= 2.83 + 0.322 + 0.238 + 0.12 + 0.197 = 3.707$ mm
Average S/N ratio = 8.91
$n_{average\ S/N} = 8.91 + (9.835 - 8.91) + (9.274 - 8.91) + (9.580 - 8.91) + (9.528 - 8.91)$
$= 8.91 + 0.925 + 0.364 + 0.67 + 0.6 = 11.5$ mm
$\% \text{ Error} = \frac{Experimental\ value - Predicted\ Value}{Predicted\ Value} * 100$
$\% \text{ Error} = \frac{3.744 - 3.6707}{3.6707} * 100 = 1.99\ \%$

4.8 Regression equation for all the response

The regression equation has been developed by using Minitab18 statistical software. The second-order polynomial regression equation representing the bead

Exp. no.	Process Parameter				1
	P	B	F	T	BW
1	140	60	50	35	8.91
2	140	70	75	40	7.65
3	140	80	100	45	6.60
4	140	90	125	50	9.38
5	140	100	150	55	8.04
6	150	60	75	45	7.19
7	150	70	100	50	9.84
8	150	80	125	55	8.21
9	150	90	150	35	7.36
10	150	100	50	40	9.88
11	160	60	100	55	8.38
12	160	70	125	35	7.43
13	140	60	50	35	10.51
14	160	90	50	45	8.59
15	160	100	75	50	7.90
16	170	60	125	40	11.11
17	170	70	150	45	8.89
18	170	80	50	50	8.43
19	170	90	75	55	10.59
20	170	100	100	35	8.78
21	180	60	150	50	7.72
22	180	70	50	55	11.25
23	180	80	75	35	9.93
24	180	90	100	40	8.49
25	180	100	125	45	11.77

Table 15.
S/N Ratio from MINITAB 18.

Level	Peak current (P)	Base current (B)	Pulse frequency (F)	Pulse on time (%)
1	8.115	8.661	9.411	8.483
2	8.495	9.013	8.652	9.528
3	8.562	8.736	8.417	8.609
4	9.559	8.881	9.580	8.653
5	9.835	9.274	8.504	9.293
Delta	1.720	0.613	1.163	1.045
Rank	1	4	2	3

Table 16.
Response table for S/N ratio.

Level	Peak current (P)	Base current (B)	Pulse frequency (F)	Pulse on time (%)
1	2.561	2.745	2.977	2.672
2	2.683	2.861	2.741	3.027
3	2.700	2.768	2.654	2.755
4	3.029	2.801	3.068	2.720
5	3.152	2.950	2.685	2.951
Delta	0.591	0.205	0.414	0.355
Rank	1	4	2	3

Table 17.
Response table for mean.

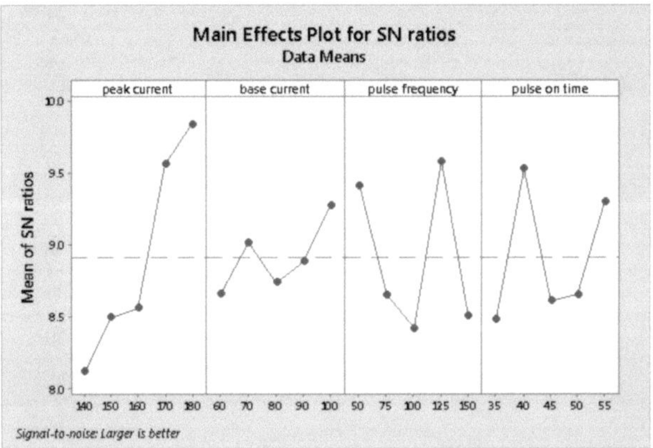

Figure 1.
Main effect plot for S/N ratio: BW.

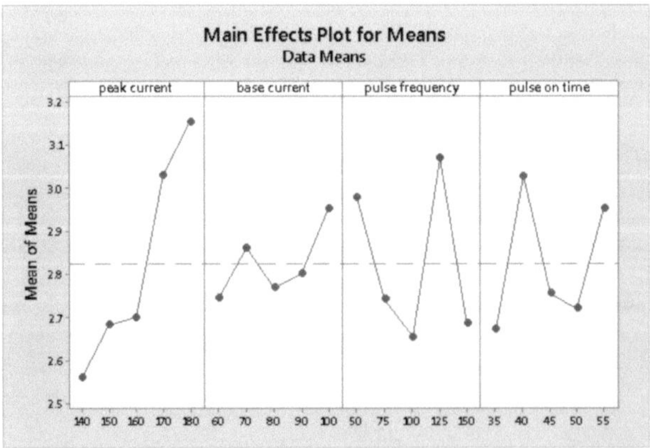

Figure 2.
Main effect plot for mean: BW.

geometry expressed as a function of peak current, base current, pulse frequency and pulse on time as given in Eq. (3). The predicted result as per the regression equation is shown in **Table 20**. After that % error between predicted value and experimental value is obtained as given in **Table 21** [10].

Parameter/control factor	Optimum parameter	Level	Optimum value
Peak current	1	5	180A
Base current	4	5	100A
Pulse frequency	2	4	125Hz
Pulse on time	3	2	40%

Table 18.
Optimum parameter for bead width.

	Prediction	Experiment				
Level	P5B5F4T2 (180A, 100A, 125Hz, 40%)	P5B5F4T2 (180A, 100A, 125Hz, 40%)				
		Exp.1	Exp.2	Exp.3	Exp.4	Exp.5
		3.75 mm	3.72 mm	3.70 mm	3.78 mm	3.72 mm
		Average				
Bead width	3.707 mm	3.744 mm				
S/N Ratio	11.5	11.86				

Table 19.
Confirmatory results for bead width.

Exp. no.	Process parameter				1
	P	B	F	T	BW
1	140	60	50	35	2.82
2	140	70	75	40	2.44
3	140	80	100	45	2.17
4	140	90	125	50	2.97
5	140	100	150	55	2.55
6	150	60	75	45	2.32
7	150	70	100	50	3.13
8	150	80	125	55	2.60
9	150	90	150	35	2.36
10	150	100	50	40	3.15
11	160	60	100	55	2.65
12	160	70	125	35	2.38
13	140	60	50	35	3.38
14	160	90	50	45	2.72
15	160	100	75	50	2.51
16	170	60	125	40	3.62
17	170	70	150	45	2.81
18	170	80	50	50	2.67
19	170	90	75	55	3.41

Exp. no.	Process parameter				1
	P	B	F	T	BW
20	170	100	100	35	2.78
21	180	60	150	50	2.46
22	180	70	50	55	3.68
23	180	80	75	35	3.17
24	180	90	100	40	2.69
25	180	100	125	45	3.91

Table 20.
Predicted result from the regression equation.

Exp. no.	Process parameter				1
	P	B	F	T	BW
1	140	60	50	35	1.1%
2	140	70	75	40	1.2%
3	140	80	100	45	1.4%
4	140	90	125	50	1.0%
5	140	100	150	55	1.2%
6	150	60	75	45	1.3%
7	150	70	100	50	1.0%
8	150	80	125	55	1.2%
9	150	90	150	35	1.3%
10	150	100	50	40	1.0%
11	160	60	100	55	1.1%
12	160	70	125	35	1.3%
13	140	60	50	35	0.9%
14	160	90	50	45	1.1%
15	160	100	75	50	1.2%
16	170	60	125	40	0.8%
17	170	70	150	45	1.1%
18	170	80	50	50	1.1%
19	170	90	75	55	0.9%
20	170	100	100	35	1.1%
21	180	60	150	50	1.2%
22	180	70	50	55	0.8%
23	180	80	75	35	1.0%
24	180	90	100	40	1.1%
25	180	100	125	45	0.8%

Table 21.
Percentage error between predicted & experimental results.

Source	DF	Adj SS	Adj MS	F	P	% contribution
Peak current	4	1.2702	0.31754	0.97	0.475	24.42 %
Base current	4	0.1357	0.03393	0.10	0.978	2.61 %
Pulse frequency	4	0.6903	0.17256	0.53	0.720	13.27 %
Pulse on time	4	0.4801	0.12002	0.37	0.827	9.23 %
Error	8	2.6249	0.32812			
Total	24	5.2012				

Table 22.
ANOVA for bead width.

$$
\begin{aligned}
\text{Bead width (mm)} = {} & 2.825 - 0.264\,\text{peak current\_140} - 0.142\,\text{peak current\_150} \\
& - 0.125\,\text{peak current\_160} + 0.204\,\text{peak current\_170} \\
& + 0.327\,\text{peak current\_180} - 0.080\,\text{base current\_60} \\
& + 0.036\,\text{base current\_70} - 0.057\,\text{base current\_80} \\
& - 0.024\,\text{base current\_90} + 0.125\,\text{base current\_100} \\
& + 0.152\,\text{pulse frequency\_50} - 0.084\,\text{pulse frequency\_75} \\
& - 0.171\,\text{pulse frequency\_100} + 0.243\,\text{pulse frequency\_125} \\
& - 0.140\,\text{pulse frequency\_150} - 0.153\,\text{pulse on time\_35} \\
& + 0.202\,\text{pulse on time\_40} - 0.070\,\text{pulse on time\_45} \\
& - 0.105\,\text{pulse on time\_50} + 0.126\,\text{pulse on time\_55}
\end{aligned} \tag{3}
$$

4.9 ANOVA for all the response

ANOVA test the hypothesis that the means of two or more population are equal. AVOVA is a computational technique to quantitatively estimate the contribution that each parameter makes on the overall observed response. By using ANOVA percentage contribution of each parameter is obtained as shown in **Table 22.**

1. Peak Current $\frac{1.2702}{5.2012}$ X 100 = 24.42 %

2. Base current $\frac{0.1357}{5.2012}$ X 100 =2.67 %

3. Pulse frequency $\frac{0.6943}{5.2012}$ X 100 =13.27 %

4. Pulse on time $\frac{0.4801}{5.2012}$ X 100 = 9.230%

5. Conclusion

In this Taguchi approach is applied to determine the most influencing process parameter which effect the output response i.e BW (Bead Width). By using Minitab 18 statistical analysis software all possible combination of all input process parameter has been established Using L_{25} orthogonal array experiment has been conducted to determine S/N ratio. The response table is developed to determine the rank of each parameter. Main effect plot obtained from Minitab 18 statistical analysis software is used to determine the most influencing process parameter and their significant level. Optimum parameter for bead width is 180A peak current, 100A base current, 125 Hz pulse frequency and 40% pulse on time. The confirmatory test has been conformed to verify the optimum result obtained. ANOVA is representing the significance of each individual parameter with their % contribution.

Acknowledgements

The authors would like to express their sincere thanks to ACMS (Advance center for material science) IIT Kanpur for providing research facility.

Author details

Asif Ahmad
PSIT-Kanpur, U.P., India

*Address all correspondence to: aauptu2015@gmail.com

References

[1] Taguchi G. System of Experimental Design. White Plains, New York, NY: Unipub/Kraus International Publication; 1987

[2] MONTGOMERY D C: Design and analysis of experiments [M]. 8th ed. New York: John Wiley and Sons, 2013.

[3] Ghani JA, Choudhury IA, Hassan HH. Application of Taguchi method in the Optimization of end milling parameters. Journal of Materials Processing Technology. 2004;**145**:84-92

[4] George, P.M, PillaiN, and Shah, N: Optimization of shot peening parameters using Taguchi technique. Journal of Materials Processing Technology, 153, pp. 925-930. (2004).

[5] Ng Chin Fei, NikMizamzul Mehat and Shahrul Kamaruddin: Practical Applications of Taguchi Method for Optimization of Processing Parameters for Plastic Injection Moulding: A Retrospective Review. Hindawi Publishing Corporation ISRN Industrial Engineering Volume 2013, Article ID 462174, 11 pages doi:10.1155/2013/462174. (2013)

[6] Asif Ahmad and S. Alam. "Grey Based Taguchi Method for Optimization of TIG Process

[7] Kumar A, Sundarrajan S. Optimization of Pulsed TIG Welding Process Parameters on Mechanical Properties of AA 5456 Aluminum Alloy Weldments. Materials and Design, April 2009. 2009;**30**(4):1288-1297

[8] Prakash Kumar Sahu and Sukhomay Pal: Multi response optimization of process parameters in friction stir welded AM20 magnesium alloy by Taguchi grey relational analysis", Journal of Magnesium and Alloys. (2015).

[9] Ugur Esme, Melih Bayramoglu, Yugut Kazancoglu and Sueda Ozgun: OPTIMIZATION OF WELD BEAD GEOMETRY IN TIG WELDING PROCESS USING GREY RELATION ANALYSIS AND TAGUCHI METHOD. Materiali in tehnologije/Materials and technology 43 (2009) 3, 143–149. ISSN 1580-2949. Original scientific article. MTAEC9, 43(3)143(2009).

[10] Sivaraos, K.R.Milkey, A.R. Samsudin, A.K.Dubey and P.Kidd: Comparison between Taguchi Method and Response Surface Methodology (RSM) in Modelling CO_2 Laser Machining. Jordan Journal of Mechanical and Industrial Engineering. Volume 8 Number 1, February. 2014 ISSN 1995-6665 Pages 35 – 42. (2014)

Chapter 11

Exponentially Weighted Moving Averages of Counting Processes When the Time between Events Is Weibull Distributed

Ross Stewart Sparks and Hossein Hazrati-Marangaloo

Abstract

There are control charts for Poisson counts, zero-inflated Poisson counts, and over dispersed Poisson counts (negative binomial counts) but nothing on counting processes when the time between events (TBEs) is Weibull distributed. In our experience the in-control distribution for time between events is often Weibull distributed in applications. Counting processes are not Poisson distributed or negative binomial distributed when the time between events is Weibull distributed. This is a gap in the literature meaning that there is no help for practitioners when this is the case. This book chapter is designed to close this gap and provide an approach that could be helpful to those applying control charts in such cases.

Keywords: average run length, counts, monitoring, time between events

1. Introduction

Statistical process control and monitoring (SPCM) methods originally arose in the context of industrial/manufacturing applications, developed during and after World War II. Since then, it has become a popular way of monitoring all processes. Today, large volumes of data are often available from a variety of sources, in a variety of environments that need to be monitored. This means one needs to make sense of these data and then be able to make efficient monitoring decisions based on them. While constructing and applying monitoring tools, a fundamental assumption, necessary to justify the end results, involves the assumption about the distribution that the data have been generated from. This is the heart of outbreak detection of events particularly for estimating the in-control false discovery rates.

Selecting an appropriate probability distribution for the data is one of the most important and challenging aspects of data analysis. The estimates of outbreak false discovery rates often hinge on this crucial selection. We focus on the distribution for the time between events (TBEs) because there are often many of these events as compared to counts. Therefore, it is easier to fit an appropriate distribution using these many TBE values. The most commonly assumed distribution in the application of TBE is the Weibull distribution which is asymmetric and sometimes severely skewed. However, depending on the context, other distributions may also be used, such as the exponential distribution (which is a special case of a Weibull distribution) or the gamma distribution. If the TBE distribution is exponentially distributed, then the related counts are Poisson distributed. This book chapter focusses on counting processes when the distribution of TBE values is known to be Weibull distributed.

The challenge of making and meeting the distributional assumption is faced by all practitioners and data analysts. In many monitoring settings, event data are collected in a nearly continuous stream, and it is often more meaningful to monitor the individual TBE data [1–4] when outbreaks are of large magnitude. This individual event data are aggregated over fixed time intervals (e.g., daily) to form counts. In this chapter, these counts are monitored to detect outbreaks resulting from small changes in the incident of events. The focus is on the steady-state situation because this is the most common situations in event monitoring. Note that we cannot stop the process and investigate the out-of-control situation because often in nonmanufacturing settings it is not under our control. Events may include warranty claims of a product, health presentation at emergency departments, sales of an online products, etc. Here the term "quality of the process" is used in a general sense, which is context dependent. In the case of sales, an outbreak would represent an increased sales opportunity, provided the inventory stock can support this outbreak and the products are not sold out before the next order arrives. However, for warrantee claims, this would represent an undesirable outbreak of increased claims which may require a failure mode effect analysis [5]. Monitoring of in-control nonhomogeneous counting processes has traditionally been carried out using either Poisson or the negative binomial distributions for the counts. Many statistical tools used in SPCM for counts data are documented in Sparks et al. [1–3], Sparks et al. [6], Weiß [7, 8], Weiß and Testik [9, 10], Yontay et al. [11], and Albarracin et al. [12]. These control charts are perhaps the most well-known count monitoring methodologies. The control chart graphic is a time series plot of a signal-to-noise ratio designed for the user to make decisions about the outbreak of events.

In any case, before defining an "in-control" process, we need the information about the probability distribution of the events being monitored. When this information is available, it is possible to calculate the probabilities (or the chance) from the event in-control distribution, which could be defined by the TBEs or the counting of events within a fixed time interval. Deciding on whether event distribution constitutes an event outbreak is based on whether the event distribution is extreme compared to what is usual, i.e., counts are higher than expected. This is usually gauged by some upper threshold for the signal-to-noise ratios.

In a vast majority of SPCM applications, it is common to assume that the underlying probability distribution is of a (given) known form. In this chapter we assume that the TBEs are distributed as a Weibull distribution. However, we explore approaches to monitoring the counts of these events over fixed periods of varying length to find the period width that leads to earlier detection of outbreaks in terms of the average time to signal (ATS). The distribution of these counts when the TBEs are Weibull distributed is neither Poisson nor negative binomial distributed. Therefore, this chapter offers a different approach compared to others in the literature. In addition, the appropriate period of aggregation for the counts is explored in terms of how it influences early detection of outbreak events.

In practice event data are often collected in a continuous stream defined by the TBEs. Besides, wherever outbreaks are of large magnitude, it is more meaningful to monitor these individual TBE values [4]. In such situations, one also needs to deal with the issue of autocorrelation which may be thought of as the effect of time, as data values in close proximity with respect to time and space are likely dependent. This violates one of the basic assumptions in process monitoring, and a common way to deal with this issue is by fitting a time series model and monitoring the standardized residuals. The assumption of a distribution is also an important part of this analysis. In this chapter, our focus is on the Weibull distributional assumption for the TBE values. However, rather than monitoring the TBEs, we monitor the counts over a fixed time interval, because this improves the early detection of smaller outbreaks (Sparks et al., [4]).

2. Monitoring homogeneous counts

Considering a fixed distribution for TBEs during the monitoring period, we assume that the time of the day and date stamp of all events are available. For a series comprised of n events, let the day numbers for events be denoted as

$$d_1, d_2, \dots, d_n$$

with the event times within days which are defined as

$$\tau_1, \tau_2, \dots, \tau_n.$$

Note that these times are measured in fractions of a day, e.g., $0 \le \tau_i < 1$. Daily counts are given by counting the number d_i values that are the same and then adding the days with zero counts for those days that are missing from the d_i values. Then TBEs are given by

$$w_1 = d_2 - d_1 + \tau_2 - \tau_1, \dots, w_{n-1} = d_n - d_{n-1} + \tau_n - \tau_{n-1}$$

where w_i represents the time between $(i+1)$th and ith events. We flag an outbreak wherever w_i for any $i = 1, \dots, n-1$ are consistently lower than an expected value. These TBEs are assumed to be Weibull distributed with fixed scale and shape parameters. Using R code, we define the counting process, for say daily counts, as

$$x < -rweibull(n = number\ of\ TBE, shape = shape, scale = scale)$$
$$counts < -table(floor(cumsum(x)))$$
$$counts < -counts[-length(counts)]$$

Denote these counts as c_i for the ith day. We define the exponentially weighted moving average (EWMA) statistic for these daily homogeneous counts as follows:

$$e_i = \max(0, \alpha c_i + (1-\alpha)e_{i-1}) \quad (1)$$

where $e_0 = E(c_i)$ is assumed constant when in-control for days $i = 1, 2, \dots$. The smoothing parameter, $\alpha(0 < \alpha < 1)$, determines the level of memory of past observations included in e_i. Smaller values of α retain the more memory of past counts than do the larger values. Therefore, smaller values of α are more efficient at detecting smaller changes in mean counts, while larger values of α are more efficient at detecting larger changes. EWMA statistic, e_i, has a minimum value of zero, and we do not allow it to fall below zero because we are only interested in outbreaks, i.e., counts that are higher than expected. This EWMA statistic controls the worst-case scenarios, whereas the traditional EWMA statistic can wonder well below zero before going out of control causing the EWMA statistic to take a long time to signal the outbreak. An outbreak is flagged when e_i exceeds a predetermined threshold, $h_c(ARL_0, scale, shape, \alpha)$. For a given pair of shape and scale parameters,

h_c is determined so that a desired in-control average run length (ARL_0) is achieved. Appendix A provides models for establishing the thresholds for various values for $0.2 \leq scale \leq 0.46$, $0.6 \leq shape \leq 1.4$, $\alpha = 0.1$ and $ARL_0 = 100, 200, 300$ or 400. Given a h_c and the process is out of control, we want the ARL to be as low as possible. Monitoring daily counts, the *ARL* is defined as the number of days before a signaled outbreak. Later we will use the ATS to assess the relative performance of control chart plans, because the ARL may vary according to the aggregation period we select for the counts. Note that a false outbreak is flagged by this EWMA statistic being significantly larger than expected given it is in-control.

3. Monitoring in-control nonhomogeneous counts

For most of the real-world cases, TBE distribution may vary due to different circumstances while not experiencing an outbreak. For instance, if we aggregate the TBEs in daily intervals, occurring intervals for the events may vary based on the day of the week, and even working and non-working days may affect the distribution of TBEs. As a result, we often face nonhomogeneous count processes. Hence, we define an adaptive exponentially weighted move average (AEWMA) statistic for nonhomogeneous daily counts as

$$ae_i = \max\left(0, \alpha c_i / h_c(ARL_0, scale, shape, \alpha) + (1-\alpha) ae_{i-1}\right) \qquad (2)$$

where ae_i is the AEWMA statistic at time t, and $ae_0 = E(c_i)/h_c(ARL_0, scale, shape, \alpha)$ for days $i = 1, 2, \ldots$. The other notations are as defined earlier. To control the false discovery rate, we set $h_c(ARL_0 = c, scale, shape, \alpha)$ so that a desiredARL_0 is achieved. An outbreak is flagged approximately whenever $ae_i > 1$.

The $\alpha(0 < \alpha < 1)$ in Eq. (1) determines the level of memory of past observations in this average ae_i. Smaller α values retain more memory of past counts; therefore, small α values are efficient at retaining enough memory of past counts to have the power to flag smaller outbreaks. However, larger values of α are needed when there is a larger size outbreak, because a shorter range of memory is adequate to build enough power to detect the outbreak. In addition, the shorter memory for the EWMA average with larger values of α is less inclined to be influenced by too many past in-control counts.

4. Simulation results

To assess the validity and applicability of our proposed adaptive method, we employ simulations studies. We restrict our attention to plans that have $10 < E(c_i) < 50$ which limits the volume of simulations that are required to make a realistic judgment. This is also the range where the EWMA counts become competitive for small changes in the scale parameter (see [4] and Sparks et al., 2020). To do so, we consider four counting processes with daily aggregations. The shape parameters for the TBE distributions when the processes are in-control are equal to 0.85, 0.95, 1.15, and 1.25. The in-control scale parameter for each process is equal to 0.02, 0.025, 0.03, and 0.035, respectively. We intentionally imposed outbreaks in the simulated data to assess if the proposed method is capable of detecting them. In the simulation study, we assume that the outbreaks result in a decrease in the scale parameter. Various surveillance plans are devised to monitor each process. The ATS_0 for the plans associated with the first process (with shape and scale parameters of 0.85 and 0.02, respectively) is set to be equal to 400. The ATS_0 for the plans associated with other processes are equal to 300, 200, and 100, respectively. 100,000 events are employed to estimate the count mean for each process. In addition, 500 events are also considered for the burn-in period of the simulations. The performance of the devised plans is presented in **Tables 1–4**. The lowest ATS values are colored in the tables below in black bold text to make it easier to see which plans are more efficient in certain situations.

Table 1 shows the performance results for the plans employed to monitor counting process where the in-control data are Weibull with scale and shape parameters of 0.035 and 1.25, respectively. ATS_0 (in days) for these plans are set approximately equal to 100. As shown in **Table 1**, for the outbreaks of larger magnitude, plans with larger smoothing parameter are superior. On the contrary,

	EWMA counts (α), shape = 1.25							
Threshold	**31.835**	**32.162**	**32.524**	**32.807**	**33.207**	**33.470**	**33.803**	**34.108**
Scale	α							
	0.04	**0.06**	**0.08**	**0.10**	**0.125**	**0.15**	**0.175**	**0.20**
0.035	100.05	101.16	100.41	101.95	100.31	100.02	100.55	101.21
0.034	41.952	40.255	**39.997**	41.552	40.954	38.936	40.372	43.733
0.033	19.116	18.757	**18.275**	19.098	19.564	19.106	19.538	20.302
0.032	13.054	11.924	11.721	**11.259**	11.359	10.733	11.370	11.843
0.031	8.598	8.231	8.156	7.860	7.847	**7.491**	7.515	7.566
0.030	6.774	5.937	5.943	5.749	5.628	**5.393**	5.508	5.437
0.029	5.824	4.715	4.724	4.696	4.426	4.211	**4.161**	4.183
0.028	4.408	4.149	3.912	3.739	3.660	3.424	3.446	**3.283**
0.027	3.699	3.358	3.386	3.025	3.002	2.886	2.832	**2.788**

Table 1.
Performance of plans when the in-control TBEs are Weibull distributed with scale = 0.035 and shape = 1.25.

	EWMA counts (α), shape = 1.15							
Threshold	**36.20**	**37.058**	**37.4917**	**37.904**	**38.315**	**38.725**	**39.1025**	**39.515**
Scale	α							
	0.04	**0.06**	**0.08**	**0.10**	**0.125**	**0.15**	**0.175**	**0.20**
0.03	205.09	200.54	201.68	201.21	200.23	200.82	200.96	200.51
0.029	50.792	**47.665**	54.813	53.764	52.916	54.227	56.616	61.987
0.028	22.027	21.034	20.747	**21.067**	21.624	22.292	23.361	23.265
0.027	12.791	11.790	11.953	11.655	**11.265**	11.344	11.397	12.319
0.026	8.678	8.414	7.811	7.414	7.211	**7.230**	7.333	7.497
0.025	6.718	6.163	5.718	5.508	5.444	5.243	**5.159**	5.162
0.024	5.078	4.650	4.462	4.283	4.228	3.921	3.920	**3.877**
0.023	4.482	4.010	3.742	3.558	3.465	3.190	3.155	**3.072**

Table 2.
Performance of plans when the in-control TBEs are Weibull distributed with scale = 0.03 and shape = 1.15.

	EWMA counts (α), shape = 0.95							
Threshold	**41.35**	**41.9028**	**42.4899**	**42.9932**	**43.5669**	**44.1448**	**44.7079**	**45.1527**
Scale				α				
	0.04	**0.06**	**0.08**	**0.10**	**0.125**	**0.15**	**0.175**	**0.20**
0.025	302.19	301.69	302.03	301.51	302.29	301.85	301.29	301.83
0.024	**51.937**	56.110	59.550	65.442	63.928	66.311	78.925	74.448
0.023	23.359	**20.844**	22.686	22.031	22.855	23.860	25.744	26.173
0.022	12.681	11.357	11.807	**11.345**	11.412	11.493	12.104	12.056
0.021	8.784	8.253	7.723	7.394	**7.150**	7.326	7.196	7.231
0.0205	7.474	6.866	6.855	6.410	5.997	**5.962**	6.113	5.977
0.020	6.788	5.962	5.416	5.368	5.075	5.038	5.136	**4.976**
0.019	4.840	4.707	4.415	4.032	3.942	3.844	3.805	**3.757**

Table 3.
Performance of plans when the in-control TBEs are Weibull distributed with scale = 0.025 and shape = 0.95.

	EWMA counts (α), shape = 0.85							
Threshold	**48.7256**	**49.5976**	**50.3141**	**50.9389**	**51.6532**	**52.35**	**53.0104**	**53.5827**
Scale				α				
	0.04	**0.06**	**0.08**	**0.10**	**0.125**	**0.15**	**0.175**	**0.20**
0.02	401.92	401.28	399.92	400.89	399.89	401.09	401.08	399.93
0.0195	**113.23**	121.88	123.74	135.49	144.80	145.77	145.82	156.46
0.019	51.948	**48.144**	51.531	57.694	57.722	63.461	66.637	69.652
0.018	19.344	18.374	17.494	**17.316**	17.663	18.892	19.627	19.487
0.0175	12.846	12.711	12.370	12.263	**12.260**	12.272	12.651	12.765
0.017	10.966	10.696	9.583	9.032	9.052	**8.752**	9.179	9.074
0.0165	8.261	8.510	7.599	7.301	6.946	6.948	**6.868**	6.954
0.016	7.191	6.386	6.198	5.803	5.666	5.429	5.456	**5.277**
0.015	5.474	4.868	4.354	4.119	4.031	4.047	3.862	**3.589**

Table 4.
Performance of plans when the in-control TBE are Weibull distributed with scale = 0.02 and shape = 0.85.

for the early detection of the small outbreaks, smaller values for the smoothing parameter are preferred.

Table 2 shows the performance results of the plans when the shape and the scale parameters of the Weibull distribution are equal to 1.15 and 0.03, respectively. The ATS_0 for the plans employed to monitor this counting process is set to be approximately 200. In the results indicated in **Table 1**, plans with smaller smoothing parameter work better in early detection of small outbreaks than those with larger smoothing parameter. For the larger outbreaks, the detection power of the plans increases as the smoothing parameters increases.

Table 3 shows the performance results of the plans when the shape and the scale parameters of the Weibull distribution are equal to 0.95 and 0.025, respectively. The ATS_0 for the plans employed to monitor this counting process is set to be approximately 300. Conclusions similar to those regarding **Tables 1** and **2** can be drawn from **Table 3**. It is clearly observed that the larger the magnitude of the outbreak, the larger the smoothing parameter should be. On the other hand, for the detection of small outbreaks, plans with larger values of ATS_0 need even smaller values for smoothing parameters than do the plans with smaller ATS_0. As shown in **Table 1**, smoothing parameters equal to or larger than 0.08 work better for outbreak detection. As the ATS_0 increases, analogous to results presented in **Tables 2** and **3**, even smaller values for α are needed to devise a plan of larger detection power. Last but not the least, similar results can be driven from **Table 4**, which presents the performance of the monitoring plans applied to a counting process with underlying Weibull distribution for TBEs. The shape and the scale parameters of the aforementioned distribution are equal to 0.85 and 0.028, respectively.

5. Real-world example

In this section, we apply our proposed method to a real-world example. The counting process to monitor is the number of presentations at Gold Coast University Hospital emergency department for a broad definition of influenza. The events

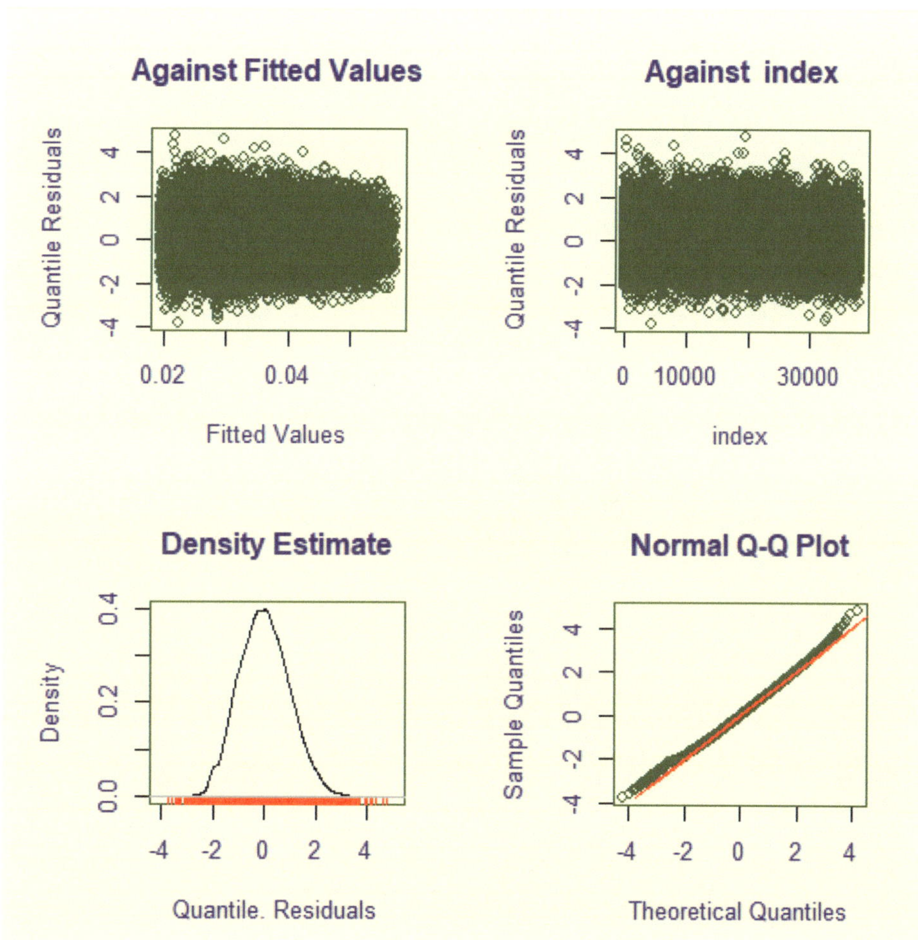

Figure 1.
Residual analysis for the fitted model.

are presentations at Gold Coast University Hospital with flu symptoms. Data is gathered for four consecutive years starting from January 2015. We first check if the TBEs are Weibull distributed. We fitted a Weibull regression model to data, and all the parameters of the conditional distribution of the response variable are modeled using explanatory variables as the hour of the day harmonics, seasonal harmonics, and day of the week. To do so, we employ "gamlss" [13] R package for statistical modeling. This package includes functions for fitting the generalized additive models for location, scale, and shape introduced by Rigby and Stasinopoulos [14]. The R procedure for the model fitting is as follows:

```
gamlss(formula = TBE ~ wd + wd=="Monday") + (wd=="Thursday"))*
(cos(2 * pi * hr/24) + sin(2 * pi * hr/24)) + cos (2 * pi * nday/365.25)
+ sin (2 * pi * nday/365.25) sigma.formula = ~(cos2 * pi * hr/24) + sin
(2*pi*hr/24))+cos(2*pi*nday/365.25)+sin(2*pi*day/365.25)
family=WEI()data=data)
```

where wd, nday, and hr. are week day, number of the day in a year, and the time of day (0–24), respectively. **Figure 1** summarizes the analysis of the residuals for the fitted model. As shown in **Figure 1**, since residuals do not represent any particular pattern in data, and the model describes the response variables quite well, then we conclude that TBEs are Weibull distributed. Details for the analyzing the model adequacy is presented in Appendix B.

As mentioned earlier, the threshold, h_c, is determined according to parameters of the underlying Weibull distribution for the TBEs. For any timestamp during the monitoring period, the fitted model is used to predict the parameters of the Weibull distribution. Then these parameters along with the smoothing parameter, α, and the in-control ARL are substituted in models in Appendix A to establish the h_c for Weibull-distributed counts. The parameter for the day of the count is taken as the average of the estimated parameters for the day of events. Considering $\alpha = 0.1$ and

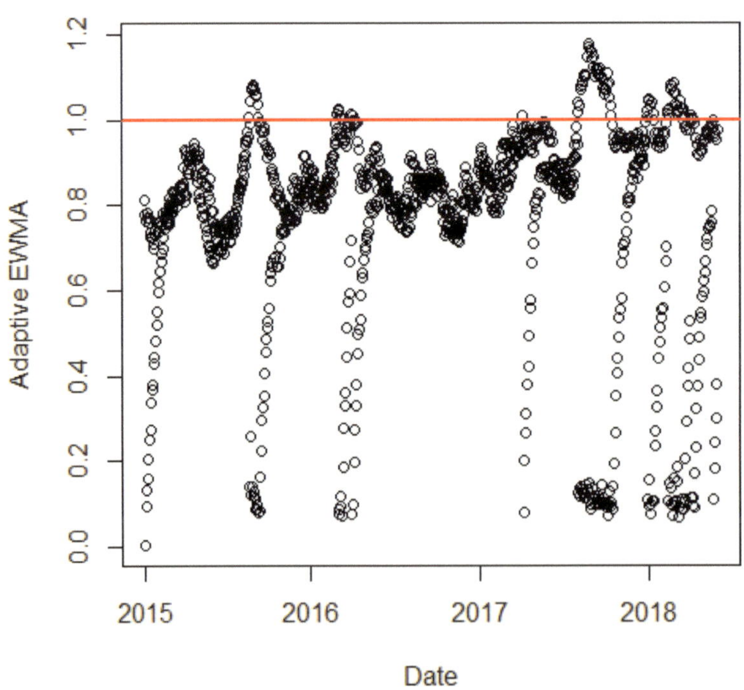

Figure 2.
AEWMA chart using $\alpha = 0.1$ for influenza presentations at Gold Coast University Hospital.

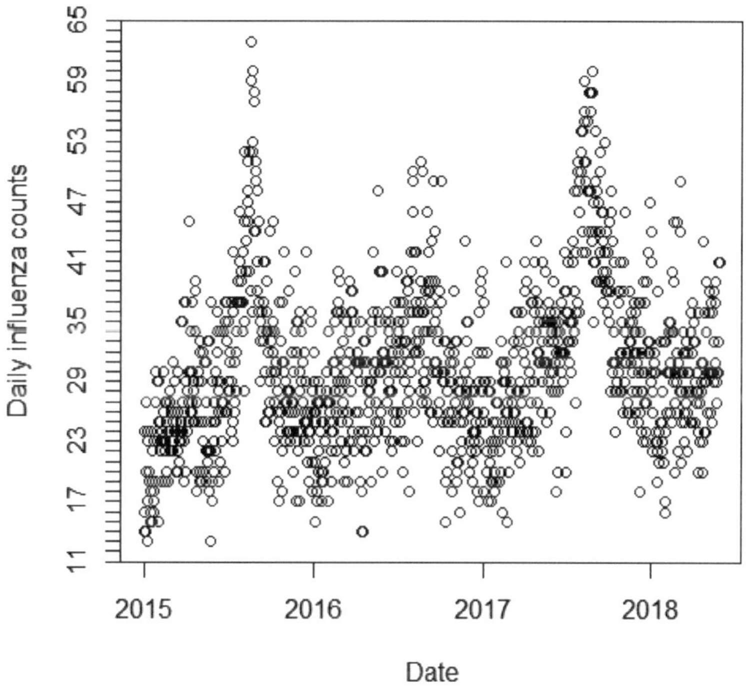

Figure 3.
Event counts vs. time.

ARL_0 for the plan to be 400, the estimated parameters are used to establish the AEWMA plan for monitoring the daily counts of flu presentations at the Gold Coast University Hospital. **Figure 2** shows the time series of the monitoring statistic over the study period, and **Figure 3** is the time series of the daily flu counts. The proposed plan indicates six potential flu outbreaks during 4 years as the monitoring statistic falls beyond 1. Note that influenza counts in 2015 were bigger than usual, but not unusual in 2016 (apart from early in that year). However, in 2017 this was flagged as a very unusual influenza outbreak and this seems to persist into early 2018.

6. Concluding remarks

In this chapter, we proposed an adaptive EWMA surveillance plan to monitor a counting process of which the time between its events is Weibull distributed. The proposed method can be applied to both homogeneous and nonhomogeneous processes. To implement the proposed surveillance plan, the scale and shape parameters for the underlying distribution of the TBEs are estimated using a distributional regression approach [15]. Then the threshold for the counts is established using the estimated parameters and the desired ARL. The proposed plan is applied to both simulated and real data. Simulation results indicate that the proposed method is applicable for detecting outbreaks of any magnitude and also signals them in a reasonable time after their incidence. In addition, simulations revealed that for the detection of the large outbreak, plans with larger smoothing parameter are superior. However, for the early detection of small outbreaks, we need to employ smaller smoothing weights. Applying the proposed surveillance method to real data, we conclude that the proposed method is capable of detecting outbreaks in nonhomogeneous counting processes.

A. Appendix

The thresholds for the linear model fitted

$$\text{sqrt(hc)} \sim (\text{scale} + \text{shape} + \log(\text{scale}) + \log(\text{shape}))^2 + \text{sqrt(shape)} * \text{scale}$$

The multiply R-squared for all these models is equal to 1 corrected to the sixth decimal place. The square of the fitted values for these models provides an starting estimate of the threshold for the respectively in-control ARL of 100, 200, 300, or 400. This can be further revised using simulation (**Table A1**).

In-control ARL	100	200	300	400
(Intercept)	55.27965623580790	14.50063556867970	35.4751774678155	1.62174314752660
scale	−626.65874427126700	−750.65355682164500	−853.0616160220110	−666.97214806267300
shape	54.99181051902600	6.91623162316955	26.9351743566771	−2.52567948128196
log(scale)	−8.64992403297571	−8.66215508213318	−8.6812026091707	−8.66460977364899
log(shape)	16.34706116414410	6.99770664515964	12.4345299969720	3.32335746029235
sqrt (shape)	−132.57740079984300	−43.82444799901730	−84.7132341475612	−21.35094544552350
scale:shape	−224.25112834314800	−263.89677049308800	−301.0663396807370	−236.64136324217100
scale:log (scale)	−69.61113437236530	−70.49847595504570	−69.8492890948244	−69.64522997112020
scale:log (shape)	−160.56646815370100	−201.15404356846600	−235.8054471494210	−174.62069894929300
shape:log (scale)	2.19016501325625	2.17689719954047	2.2105503473413	2.20073389465297
shape:log (shape)	−7.97657987867648	4.93365063336234	0.1464213924397	6.88699339675467
log(scale): log(shape)	−3.19688371718715	−3.19951671299892	−3.1780225086612	−3.20431683519426
scale:sqrt (shape)	795.53717813564600	957.71276755290300	1098.4086506150800	848.24798318601300

Table A1.
Fitted models for predicting the thresholds when the scale parameter is between 0.02 and 0.046 (inclusive) and the shape parameter ranges from 0.6 to 1.4 (inclusive) for the EWMA statistic used in this chapter with smoothing constant α = 0.1.

B. Appendix

```
*******************************************************************
Family:  c("WEI", "Weibull")

Call:  gamlss(formula = TBE ~ wd + (wd == "Thursday") * (cos(2 *
    pi * hr/24) + sin(2 * pi * hr/24)) + cos(2 * pi *
    day/365.25) + sin(2 * pi * day/365.25), sigma.formula = ~(cos(2 *
    pi * hr/24) + sin(2 * pi * hr/24)) + cos(2 * pi *
    day/365.25) + sin(2 * pi * day/365.25), family = WEI(),       data = temp)

Fitting method: RS()
```

```
------------------------------------------------------------------
Mu link function:  log
Mu Coefficients:
                                          Estimate Std. Error  t value Pr(>|t|)
(Intercept)                              -3.369757   0.014082 -239.291  < 2e-16 ***
wdMonday                                 -0.078713   0.019199   -4.100 4.14e-05 ***
wdSaturday                                0.002747   0.019667    0.140   0.8889
wdSunday                                 -0.077758   0.019225   -4.045 5.25e-05 ***
wdThursday                               -0.021966   0.019891   -1.104   0.2695
wdTuesday                                -0.037717   0.019439   -1.940   0.0524 .
 wdWednesday                              -0.031362   0.019439   -1.613   0.1067
cos(2 * pi * hr/24)                       0.170919   0.008217   20.802  < 2e-16 ***
sin(2 * pi * hr/24)                       0.244262   0.008684   28.127  < 2e-16 ***
cos(2 * pi * day/365.25)                  0.143229   0.007876   18.186  < 2e-16 ***
sin(2 * pi * day/365.25)                  0.130402   0.007559   17.252  < 2e-16 ***
wd == "Thursday"TRUE:cos(2 * pi * hr/24)  0.035362   0.020530    1.722   0.0850 .
wd == "Thursday"TRUE:sin(2 * pi * hr/24)  0.030439   0.021941    1.387   0.1654
---
Signif. codes:  0 '***' 0.001 '**' 0.01 '*' 0.05 '.' 0.1 ' ' 1

------------------------------------------------------------------
Sigma link function:  log
Sigma Coefficients:
                          Estimate Std. Error t value Pr(>|t|)
(Intercept)              -0.007815   0.004079  -1.916  0.05535 .
cos(2 * pi * hr/24)      -0.028340   0.005434  -5.216 1.84e-07 ***
sin(2 * pi * hr/24)      -0.008343   0.006060  -1.377  0.16859
cos(2 * pi * day/365.25) -0.009082   0.005691  -1.596  0.11058
sin(2 * pi * day/365.25)  0.015188   0.005507   2.758  0.00581 **
---
Signif. codes:  0 '***' 0.001 '**' 0.01 '*' 0.05 '.' 0.1 ' ' 1

------------------------------------------------------------------
No. of observations in the fit:  38143
Degrees of Freedom for the fit:  18
      Residual Deg. of Freedom:  38125
                      at cycle:  5

 Global Deviance:     -187355.7
            AIC:     -187319.7
            SBC:     -187165.8
```

Author details

Ross Stewart Sparks* and Hossein Hazrati-Marangaloo
Data61, CSIRO, Sydney, NSW, Australia

*Address all correspondence to: ross.sparks@data61.csiro.au

References

[1] Sparks R, Keighley T, Muscatello D. Exponentially weighted moving average plans for detecting unusual negative binomial counts. IIE Transactions. 2010; **42**(10):721-733

[2] Sparks R, Carter C, Graham P, Muscatello D, Churches T, Kaldor J, et al. Understanding sources of variation in syndromic surveillance for early warning of natural or intentional disease outbreaks. IIE Transactions. 2010;**42**(9): 613-631

[3] Sparks RS, Keighley T, Muscatello D. Early warning CUSUM plans for surveillance of negative binomial daily disease counts. Journal of Applied Statistics. 2010;**37**(11):1911-1929

[4] Sparks R, Jin B, Karimi S, Paris C, MacIntyre CR. Real-time monitoring of events applied to syndromic surveillance. Quality Engineering. 2019; **31**(1):73-90

[5] Liu HC, Liu L, Liu N. Risk evaluation approaches in failure mode and effects analysis: A literature review. Expert Systems with Applications. 2013;**40**(2): 828-838

[6] Sparks RS, Keighley T, Muscatello D. Optimal exponentially weighted moving average (EWMA) plans for detecting seasonal epidemics when faced with non-homogeneous negative binomial counts. Journal of Applied Statistics. 2011;**38**(10):2165-2181

[7] Weiß CH. EWMA monitoring of correlated processes of Poisson counts. Quality Technology and Quantitative Management. 2009;**6**(2):137-153

[8] Weiß CH. Detecting mean increases in Poisson INAR (1) processes with EWMA control charts. Journal of Applied Statistics. 2011;**38**(2):383-398

[9] Weiss CH, Testik MC. CUSUM monitoring of first-order integer-valued autoregressive processes of Poisson counts. Journal of Quality Technology. 2009;**41**(4):389-400

[10] Weiß CH, Testik MC. The Poisson INAR (1) CUSUM chart under overdispersion and estimation error. IIE Transactions. 2011;**43**(11):805-818

[11] Yontay P, Weiß CH, Testik MC, Pelin Bayindir Z. A two-sided cumulative sum chart for first-order integer-valued autoregressive processes of poisson counts. Quality and Reliability Engineering International. 2013;**29**(1):33-42

[12] Albarracin OYE, Alencar AP, Lee Ho L. CUSUM chart to monitor autocorrelated counts using negative binomial GARMA model. Statistical Methods in Medical Research. 2018; **27**(9):2859-2871

[13] gamlss: Generalised Additive Models for Location Scale and Shape. Available from: https://cran.r-project.org/web/packages/gamlss

[14] Rigby RA, Stasinopoulos DM. Generalized additive models for location, scale and shape. JRSS: Series C (Applied Statistics). 2005;**54**:507-554. DOI: 10.1111/j.1467-9876.2005.00510.x

[15] Sparks RS. CUSUM charts for signalling varying location shifts. Journal of Quality Technology. 2000; **32**(2):157-171

Chapter 12

Simulation-Based Comparative Analysis of Nonparametric Control Charts with Runs-Type Rules

Ioannis S. Triantafyllou

Abstract

In this chapter, we study well-known distribution-free Shewhart-type monitoring schemes based on order statistics. In order to empower the in- and out-of-control performance of the control charts being under consideration, several runs-type rules are enhanced. The simulation-based experimentation carried out reveals that the proposed schemes achieve remarkable efficiency for detecting possible shifts in the distribution of the underlying process.

Keywords: distribution-free monitoring schemes, average run length, false alarm rate, Lehmann alternatives, statistical process control, order statistics

1. Introduction

Statistical process control is widely applied to monitor the quality of a production process, where no matter of how thoroughly it is maintained, a natural variability always exists. Control charts help the practitioners to identify assignable causes so that the state of statistical control can be accomplished. Generally speaking, when a cast-off shift in the process takes place, a control chart should detect it as quickly as possible and produce an out-of-control signal.

Shewhart-type control charts were introduced in the early work of Shewhart [1], and since then, several modifications have been established and studied in detail. For a thorough study on statistical process control, the interested reader is referred to the classical textbooks of Montgomery [2] or Qiu [3]. Most of the monitoring schemes are distribution-based procedures, even though this assumption is not always realized in practice. To overcome this obstruction without disrupting the primary formation of the traditional control charts, several nonparametric (or distribution-free) monitoring schemes have been proposed in the literature. The plotted statistics being utilized for constructing such type of control charts are related to well-known nonparametric testing procedures. Among others, a variety of distribution-free control charts appeared already in the literature are based on order statistics; see, e.g., Chakraborti et al. [4], Balakrishnan et al. [5], or Triantafyllou [6, 7]. For an up-to-date account on nonparametric statistical process control, the reader is referred to the review chapter of Koutras and Triantafyllou [8], the recent monograph of Chakraborti and Graham [9], or Qiu [10, 11].

In the present chapter, we study well-known distribution-free *Shewhart*-type monitoring schemes based on order statistics. The general setup of the control charts being in consideration is presented in Section 2, while their performance characteristics are investigated based on the algorithm described in Section 3. In order to enhance the ability of the proposed monitoring schemes for detecting possible shifts of the process distribution, some well-known runs-type rules are considered. In Section 4, we carry out extensive simulation-based numerical comparisons that reveal that the underlying control charts outperform the existing ones under several out-of-control scenarios.

2. The setup of well-known nonparametric control charts based on order statistics

Let us assume that a reference random sample of size m, say $X_1, X_2, \dots, X_m$, is drawn independently from an unknown continuous distribution F, namely, when the process is in-control. The control limits of the distribution-free monitoring scheme are determined by exploiting specific order statistics of the reference sample. In the sequel, test samples are drawn independently of each other (and also of the reference sample) from a continuous distribution G, and the decision whether the process is still in-control or not rests on suitably chosen test statistics. The framework for constructing nonparametric control charts based on order statistics calls for the following step-by-step procedure.

Step 1. Draw a reference sample of size m, namely, $X_1, X_2, \dots, X_m$, from the process when it is known to be in-control.

Step 2. Form an interval by choosing appropriately a pair of order statistics from the reference sample (say, e.g., (X_a, X_b), where $1 \leq a < b \leq m$).

Step 3. Draw independently future (test) samples of size n, namely, $Y_1, Y_2, \dots, Y_n$, from the underlying process.

Step 4. Pick out l order statistics $(0 < l \leq n)$ from each test sample.

Step 5. Determine the number of observations of each test sample, say R that lie between the limits of the interval (X_a, X_b).

Step 6. Configure the signaling rule by utilizing both the statistics R and the l ordered test sample observations as monitoring statistics.

The implementation of the above mechanism does not require the assumption of any specific probability distribution for the underlying process (measurements). The reference sample (usually of large size) is drawn from the underlying in-control process, while test (Phase II) samples are picked out from the future process in order to decide whether the process remains in-control or it has shifted to an out-of-control state. The proposed monitoring scheme is likely to possess the robustness feature of standard nonparametric procedures and is, consequently, less likely to be affected by outliers or the presence of skewed or heavy-tailed distributions for the underlying populations.

It is straightforward that the proposed framework requires the construction of more than one control charts, which monitor simultaneously the underlying process. In fact, the design parameter l is connected to the number of the control charts which are needed to be built for trading on the aforementioned mechanism. Indeed, for each one of the l order statistics from the test sample, a separate two-sided control chart should be constructed.

The family of distribution-free monitoring schemes presented earlier includes as special cases some nonparametric control charts, which have been already established in the literature. For example, the monitoring scheme established by Balakrishnan et al. [5] calls for the following plotted statistics:

- A quantile $Y_{j:n}$ of the test sample which is compared with the control limits (X_a, X_b)

- The number of observations from the test sample that lie between the control limits

It goes without saying that the control chart introduced by Balakrishnan et al. [5] (*Chart* 1, hereafter) belongs to the family of monitoring schemes described previously. In fact, the *BTK* chart could be seen as a special case of the aforementioned class of distribution-free control schemes with $l = 1$. According to *Chart* 1, the process is declared to be in-control, if the following conditions hold true

$$X_{a:m} \leq Y_{j:n} \leq X_{b:m} \quad \text{and} \quad R \geq r, \tag{1}$$

where r is a positive integer.

In addition, the monitoring scheme introduced by Triantafyllou [6] (*Chart* 2, hereafter) takes into account the location of two order statistics of the test sample drawn from the process along with the number of its observations between the control limits. In other words, the aforementioned control chart could be viewed as a member of the general class of nonparametric monitoring schemes with $l = 2$. According to *Chart* 2, the process is declared to be in-control, if the following conditions hold true

$$X_{a:m} \leq Y_{j:n} \leq Y_{k:n} \leq X_{b:m} \quad \text{and} \quad R \geq r, \tag{2}$$

where r is a positive integer.

In a slightly different framework, Triantafyllou [7] proposed a distribution-free control chart based on order statistics (*Chart* 3, hereafter) by taking advantage of the position of single ordered observations from both test and reference sample. More precisely, *Chart* 3 asks for an order statistic of each test sample (say $Y_{j:n}$) to be enveloped by two prespecified observations $X_{a:m}$ and $X_{b:m}$ of the reference sample, while at the same time an ordered observation of the reference sample (say $X_{i:m}$) be enclosed by two predetermined values of the test sample $(Y_{c:n}, Y_{d:n})$. *Chart* 3 makes use of an in-control rule, which embraces the following three conditions:

Condition 1. The statistic $Y_{j:n}$ of the test sample should lie between the observations $X_{a:m}$ and $X_{b:m}$ of the reference sample, namely, $X_{a:m} \leq Y_{j:n} \leq X_{b:m}$.

Condition 2. The interval $(Y_{c:n}, Y_{d:n})$ formulated by two appropriately chosen order statistics of the test sample should enclose the value $X_{i:m}$ of the reference sample, namely, $Y_{c:n} \leq X_{i:m} \leq Y_{d:n}$.

Condition 3. The number of observations of the Y-sample that are placed enclosed by the observations $X_{a:m}$ and $X_{b:m}$ should be equal to or more than r, namely, $R \geq r$.

3. The simulation procedure and some results

In the present section, we describe the step-by-step procedure which has been followed in order to determine the basic performance characteristics of monitoring schemes mentioned previously. Two well-known runs-type rules are implemented in order to improve the performance of the control charts being considered. More precisely, if we denote by *LCL* and *UCL* the lower and the upper control limit of the underlying monitoring scheme, we apply the following runs rules

- The 2-of-2 rule. Under this scenario, an out-of-control signal is produced from the control chart, whenever two consecutive plotted points fall all of them either on or above the UCL or all of them fall on or below the LCL (see, e.g., Klein [12]).

- The 2-of-3 rule. Under this scenario, an out-of-control signal is produced from the control chart, whenever two out of three consecutive plotted points fall outside the control limits (LCL, UCL) of the corresponding scheme.

We next illustrate the detailed procedure for determining the performance of *Chart* 2 enhanced with the 2-of-3 rule. It goes without saying that a similar algorithm has been constructed in order to study the corresponding characteristics of the remaining control schemes, namely, *Chart* 1 and *Chart* 3 enhanced with either the 2-of-2 or the 2-of-3 runs rule.

Step 1. Generate a reference sample of size m from the in-control distribution F and k_2 test samples of size n from the out-of-control distribution G.

Step 2. Determine the control limits of the monitoring scheme *Chart* 2, by selecting appropriately the parameters a, b, r.

Step 3. Calculate the test statistics Y_j, Y_k, R for each test sample, and examine whether *Chart* 2 produces an out-of-control signal or not, namely, whether at least one of the conditions mentioned in (2) is violated.

Step 4. Define a dummy variable $T_i,\ i = 1, 2, \dots, k_2$ for each test sample separately. The variable T_i takes on the value 0 when all conditions in (2) are satisfied, while it takes on the value 1 otherwise.

Step 5. Determine all consecutive (uninterrupted) triplets consisting of $T_i's$ elements, namely, all triplets $(T_i, T_{i+1}, T_{i+2}), i = 1, \dots, k_2 - 2$. Define the dummy variable $D_j,\ j = 1, 2, \dots, k_2 - 2$ for each triplet separately. The variable D_j takes on the value 0 when the triplet consists of at least two 0s, while it takes on the value 1 otherwise.

Step 6. Calculate the alarm rate of the monitoring scheme as $AR = \sum_{j=1}^{k_2-2} D_j / (k_2 - 2)$. When $F = G$, the aforementioned probability indicates the false alarm rate of the monitoring scheme, while in case of different distributions F, G the AR corresponds to its out-of-control alarm rate.

Step 7. Define a variable $RL_h, h = 1, 2, \dots, H$ which counts the number of $D_j's$ elements, till the first appearance of a D_j equal to 1. The so-called average run length of the monitoring scheme is calculated as $ARL = \sum_{h=1}^{H} RL_h / H$. When $F = G$ $(F \neq G)$, the aforementioned quantity indicates the in-control (out-of-control) average run length of the monitoring scheme.

All steps 1–7 are repeated k_1 times and the performance characteristics of the proposed *Chart* 2 enhanced with 2-of-3 runs rule, namely, the false alarm rate (FAR, hereafter), the out-of-control alarm rate (AR_{out}, hereafter), the in-control average run length (ARL_{in}, hereafter), and the out-of-control run length (ARL_{out}, hereafter) are estimated as the mean values of the corresponding k_1 results produced by steps 6 and 7, respectively.

In order to ascertain the validity of the proposed simulation procedure described above, we shall first apply the algorithm without embodying any runs-type rule and compare the simulation-based outcomes to the corresponding results produced by the aid of the theoretical approximation appeared in Triantafyllou [6]. The simulation study has been accomplished based on the R software environment and involves 10.000 replications. **Table 1** displays several designs of the monitoring scheme mentioned as *Chart* 2 with a nominal level of in-control performance. Since we consider the same designs as those presented by Triantafyllou [6], the exact FARs have been taken from his **Table 1**. As it is easily observed, the simulation-based results seem to be quite close to the exact values in all cases considered. For example, let us assume that we draw a reference sample of size $m = 60$ and test samples of size $n = 5$. In order to achieve a prespecified in-control performance level, namely, *FAR* equal to 1%, the remaining parameters are determined as $a = 1$, $j = 2$, $k = 4$, and $r = 1$. Under the aforementioned design, Triantafyllou [6] computed the exact *FAR* equal to 0.0096, while the simulation-based procedure proposed in the present chapter gives a corresponding *FAR* value equal to 0.0116.

A different approach for appraising the ability of a monitoring scheme to detect a possible shift in the underlying distribution is based on its run length. We next focus on the waiting time random variable N, which corresponds to the amount of random test samples up to getting the first out-of-control signal from the monitoring scheme, in order to evaluate its performance. **Table 2** displays the exact and the simulation-based average run length for several designs of *Chart 2* that meet a desired nominal level of in-control performance. The exact values of *ARL* needed for building up **Table 2**, have been picked up from Triantafyllou [6] and more specifically from **Table 2** therein.

As it is readily observed, the simulation-based results seem to be quite close to the exact values in all cases considered. For example, let us assume that we draw a reference sample of size $m = 400$ and test samples of size $n = 5$. In order to achieve a prespecified in-control performance level, namely, ARL_{in} equal to 370, the remaining parameters are determined as $a = 5$, $b = 379$, $j = 2$, $k = 3$, and $r = 2$. Under

			Reference	**sample**		
		40			**60**	
n	(a, j, k, r)	*Exact FAR*	*Simulated FAR*	(a, j, k, r)	*Exact FAR*	*Simulated FAR*
5	(1, 2, 3, 3)	0.0114	0.0119	(1, 2, 4, 1)	0.0096	0.0116
	(3, 2, 3, 2)	0.0615	0.0632	(4, 3, 4, 2)	0.0478	0.0507
	(4, 3, 4, 2)	0.0999	0.0897	(6, 3, 4, 2)	0.0969	0.1044
11	(2, 4, 8, 4)	0.0116	0.0158	(4, 4, 7, 5)	0.0108	0.0098
	(1, 2, 4, 5)	0.0431	0.0551	(7, 4, 7, 4)	0.0518	0.0573
	(1, 6, 10, 5)	0.0432	0.0437	(6, 3, 6, 5)	0.1030	0.1084
25	(4, 8, 12, 4)	0.0139	0.0150	(9, 10, 14, 5)	0.0097	0.0100
	(5, 14, 17, 5)	0.0137	0.0126	(17, 12, 14, 5)	0.1031	0.1131
	(10, 11, 14, 4)	0.0928	0.1049	(15, 11, 15, 4)	0.1037	0.1130
			Reference	**sample**		
		100			**200**	
n	(a, j, k, r)	*Exact FAR*	*Simulated FAR*	(a, j, k, r)	*Exact FAR*	*Simulated FAR*
5	(3, 3, 4, 3)	0.0119	0.0105	(7, 2, 3, 2)	0.0132	0.0159
	(5, 2, 4, 3)	0.0505	0.0566	(10, 2, 4, 2)	0.0479	0.0473
	(1, 1, 5, 3)	0.0934	0.1076	(21, 3, 4, 2)	0.1008	0.0937
11	(4, 3, 8, 5)	0.0146	0.0107	(8, 3, 6, 4)	0.0105	0.0103
	(7, 3, 7, 6)	0.0460	0.0506	(16, 4, 6, 3)	0.0106	0.0097
	(7, 3, 8, 5)	0.0524	0.0603	(26, 4, 6, 3)	0.0498	0.0492
25	(19, 11, 14, 4)	0.0113	0.0122	(39, 11, 14, 4)	0.0093	0.0090
	(24, 11, 14, 4)	0.0490	0.0463	(45, 13, 16, 6)	0.0479	0.0445
	(19, 14, 17, 5)	0.0504	0.0518	(45, 14, 17, 4)	0.0995	0.0992

Table 1.
Exact and simulation-based FAR *for a given design of* Chart 2.

ARL_0	m	n	(a, b)	(j, k, r)	Exact ARL_{in}	Simulated ARL_{in}
370	200	5	(2, 188)	(2, 3, 2)	376.4	369.9
		11	(3, 180)	(2, 3, 3)	367.7	377.7
		25	(14, 178)	(6, 9, 14)	369.9	369.2
	300	5	(5, 287)	(2, 3, 2)	358.9	365.0
		11	(19, 285)	(4, 7, 4)	372.6	374.8
		25	(20, 213)	(6, 9, 7)	367.2	373.8
	400	5	(5, 379)	(2, 3, 2)	378.7	379.1
		11	(12, 366)	(3, 6, 4)	384.8	378.4
		25	(33, 296)	(7, 10, 9)	373.0	377.9
	500	5	(6, 473)	(2, 3, 2)	369.8	372.6
		11	(15, 440)	(3, 5, 4)	369.1	374.8
		25	(32, 362)	(6, 9, 6)	369.4	365.1

Table 2.
Exact and simulation-based ARL_{in} *for a given design of* Chart 2.

the aforementioned design, Triantafyllou [6] computed the exact ARL_{in} equal to 378.7, while the simulation-based procedure proposed in the present chapter gives a corresponding ARL_{in} value equal to 379.1.

We next focus on the ability of the distribution-free monitoring scheme defined in (1), under the assumption that the process has shifted to an out-of-control state. When the process has shifted from distribution F to G, then the ability of the scheme to detect the underlying alteration is associated with the function $G \circ F^{-1}$. For example, under the well-known Lehmann-type alternative (see, e.g., van der Laan and Chakraborti [13]), the out-of-control distribution function can be expressed as $G = F^{\gamma}$, where $\gamma > 0$. **Table 3** sheds light on the out-of-control performance of *Chart* 2 by offering the corresponding alarm rate of the proposed scheme under the Lehmann alternatives with parameter $\gamma = 0.2, 10$. Since we consider the same designs as those presented by Triantafyllou [6], the exact values of AR_{out} have been copied from his **Table 3**, while the simulated results have been produced by following the procedure described earlier.

			Reference	**sample**		
		40			60	
n	(a, j, k, r)	Exact AR_{out}	Simulated AR_{out}	(a, j, k, r)	Exact AR_{out}	Simulated AR_{out}
5	(2, 2, 4, 3)	0.8284 0.5448	0.8209 0.5244	(2, 2, 4, 4)	0.7890 0.3931	0.7783 0.3589
	(3, 2, 4, 2)	0.8853 0.7351	0.8758 0.7403	(4, 2, 4, 3)	0.8800 0.7161	0.8805 0.7095
11	(2, 4, 8, 4)	0.8883 0.5357	0.8825 0.5502	(7, 4, 7, 4)	0.9812 0.9076	0.9737 0.9090
	(3, 5, 8, 4)	0.8483 0.7556	0.8424 0.7612	(6, 5, 9, 4)	0.9135 0.9678	0.9278 0.9614
25	(9, 11, 14, 4)	0.9983 0.9978	0.9978 0.9978	(12, 12, 14, 4)	0.9938 0.9982	0.9922 0.9970
	(10, 11, 14, 4)	0.9991 0.9993	0.9991 0.9994	(9, 10, 14, 5)	0.9971 0.9759	0.9929 0.9763

		Reference		sample		
		100			200	
n	(a, j, k, r)	Exact AR_{out}	Simulated AR_{out}	(a, j, k, r)	Exact AR_{out}	Simulated AR_{out}
5	(5, 2, 4, 3)	0.8530 0.6062	0.8513 0.5815	(10, 2, 4, 2)	0.8609 0.6331	0.8661 0.6414
	(3, 3, 4, 3)	0.4783 0.3333	0.4854 0.3867	(15, 2, 4, 2)	0.9056 0.8300	0.8908 0.8237
11	(7, 3, 8, 5)	0.9888 0.8133	0.9693 0.8059	(25, 5, 8, 4)	0.9527 0.9936	0.9573 0.9912
	(19, 5, 7, 4)	0.9812 0.9982	0.9768 0.9831	(26, 4, 8, 4)	0.9895 0.9953	0.9835 0.9913
25	(19, 11, 14, 4)	0.9982 0.9994	0.9974 0.9962	(39, 11, 14, 4)	0.9989 0.9999	0.9963 0.9963
	(24, 11, 14, 4)	0.9996 0.9999	0.9989 0.9989	(45, 13, 16, 6)	0.9942 0.9999	0.9963 0.9962

Each cell contains the AR's attained for Y = 0.2 (upper entry) and Y = 10 (lower entry).

Table 3.
Exact and simulation-based AR_{out} *for a given design of* Chart 2.

Each cell contains the *AR*s attained for $\gamma = 0.2$ (upper entry) and $\gamma = 10$ (lower entry).

Based on the above table, it is evident that the proposed simulation algorithm seems to come to an agreement with the corresponding exact values of the out-of-control alarm rate of *Chart* 2. For example, for a design $(a, j, k, r) = (10, 2, 4, 2)$ with reference sample size $m = 200$ and test sample size $n = 5$, the exact alarm rate for a shift to Lehmann alternative with parameter $\gamma = 0.2$ (10) equals to 86.09% (66.31%), while the simulation-based alarm rate of *Chart* 2 is quite close to the exact one, namely, it equals to 86.61% (64.14%).

4. The proposed control charts enhanced with runs-type rules

In this section, we carry out an extensive numerical experimentation to appraise the ability of the distribution-free monitoring schemes *Chart* 1, *Chart* 2, and *Chart* 3

			Chart 1					*Chart* 1 with 2-of-3 runs rule				
ARL_0	m	n	(a, b)	j	r	Exact ARL_{in}	ARL_{out}	(a, b)	j	r	Exact ARL_{in}	ARL_{out}
370	100	5	(2, 96)	2	2	363.7	15.77	(3, 71)	2	2	378.01	11.73
		11	(6, 83)	5	5	363.06	6.76	(8, 71)	4	5	364.79	6.59
	500	5	(7, 473)	3	3	382.01	10.88	(9, 352)	2	2	380.5	10.27
		11	(29, 444)	4	5	373.14	16.71	(17, 303)	4	4	360.9	9.29
500	100	5	(3, 96)	3	3	499.88	14.94	(2, 71)	2	2	497.66	12.69
		11	(6, 84)	5	5	485.85	7.79	(4, 70)	3	6	511.12	7.13
	500	5	(7, 476)	3	3	503.77	12.62	(8, 352)	2	2	488.61	11.56
		11	(28, 454)	4	5	499.79	27.63	(17, 309)	4	4	489.45	5.62

Underlying distributions: Exponential with mean equal to 2 (in-control) and 1 (out-of-control) respectively.

Table 4.
Comparison of the ARL_{out}*s with the same* ARL_{in} *for* Chart 1.

enhanced with runs-type rules for detecting possible shifts of the underlying distribution. The computations have been made by the aid of the simulation procedure presented in Section 3. **Tables 4** and **5** display the improved out-of-control performance of *Chart* 1, when the 2-of-3 runs-type rule is activated. We first compare the performance of the control charts by using a common ARL_{in} and then evaluating the respective ARL_{out} for specific shifts. Consider the case of a process with underlying in-control exponential distribution with mean equal to 2 and out-of-control exponential distribution with mean equal to 1. In **Table 4**, we present the ARL_{out} values of *Chart* 1 and the proposed *Chart* 1 enhanced with 2-of-3 runs-type rule, for $ARL_{in} = 370, 500$, $m = 100, 500$, and $n = 5, 11$. The remaining design parameters a, b, j, r, were determined appropriately, so that ARL_{in} takes on a value as close to the nominal level as possible. It is evident that the proposed monitoring scheme performs better than the one established by Balakrishnan et al. [5] for all cases considered. The fact that the ARL_{out}s that exhibit *Chart* 1 with 2-of-3 runs-type rule

		***Chart* 1**	***Chart* 1 with 2-of-3 runs rule**
θ	δ	***ARL***	***ARL***
0	1	499.88	517.21
0.25	1	562.33	127.89
0.5	1	386.09	26.63
1	1	61.22	3.83
1.5	1	8.01	1.44
2	1	2.23	1.06
0.25	1.25	49.69	35.07
0.5	1.25	39.96	13.72
1	1.25	14.83	3.85
1.5	1.25	4.76	1.66
2	1.25	2.08	1.14
0.25	1.5	13.91	17.77
0.5	1.5	12.32	9.43
1	1.5	7.07	3.65
1.5	1.5	3.53	1.85
2	1.5	1.98	1.26
0.25	1.75	6.55	10.24
0.5	1.75	6.11	7.64
1	1.75	4.40	3.49
1.5	1.75	2.83	1.92
2	1.75	1.88	1.35
0.25	2	4.04	6.93
0.5	2	3.87	5.49
1	2	3.17	3.09
1.5	2	2.37	2.03
2	2	1.77	1.45

Table 5.
Comparison of the ARL_{out}s *with the same* ARL_{in} *for* Chart 1 *under normal distribution* (θ, δ).

					Chart 2					**Chart 2 with 2-of-2 runs rule**		
FAR	***m***	***n***	***(a, b)***	***(j, k)***	***r***	***Exact FAR***	AR_{out}	***(a, b)***	***(j, k)***	***r***	***Exact FAR***	AR_{out}
0.01	50	11	(3, 45)	(4, 6)	4	0.0103	0.4575 0.9239	(14, 35)	(4, 6)	1	0.0125	0.5527 0.9463
		25	(4, 45)	(7, 9)	7	0.0131	0.7756 0.9977	(11, 40)	(7, 9)	1	0.0110	0.9194 0.9989
	100	11	(4, 92)	(3, 5)	3	0.0126	0.5832 0.9661	(23, 75)	(3, 5)	1	0.0124	0.6771 0.9782
		25	(6, 65)	(6, 8)	4	0.0092	0.8181 0.9992	(20, 90)	(6, 8)	1	0.0135	0.9563 0.9992
	500	11	(19, 405)	(3, 5)	4	0.0099	0.5976 0.9736	(122, 350)	(3, 5)	1	0.0130	0.7278 0.9846
		25	(61, 420)	(8, 10)	5	0.0092	0.9023 0.9998	(130, 430)	(8, 10)	1	0.0089	0.9551 0.9998
	1000	11	(80, 894)	(4, 6)	5	0.0101	0.6106 0.9715	(300, 550)	(4, 6)	1	0.0110	0.6234 0.9715
		25	(125, 810)	(8, 10)	8	0.0098	0.9125 0.9998	(27, 800)	(8, 10)	1	0.0094	0.9644 0.9999
0.005	50	11	(2, 45)	(4, 6)	4	0.0054	0.3250 0.8627	(2, 45)	(4, 6)	4	0.0056	0.4229 0.9990
		25	(4, 44)	(8, 10)	8	0.0049	0.6577 0.9935	(4, 44)	(8, 10)	8	0.0057	0.8633 0.9979
	100	11	(2, 83)	(3, 5)	3	0.0046	0.7508 0.9983	(2, 83)	(3, 5)	3	0.0055	0.7620 0.9983
		25	(5, 88)	(6, 8)	4	0.0044	0.7508 0.9983	(5, 88)	(6, 8)	4	0.0054	0.9169 0.9989
	500	11	(15, 420)	(3, 5)	4	0.0050	0.5234 0.9618	(15, 420)	(3, 5)	4	0.0054	0.6611 0.9799
		25	(54, 380)	(8, 10)	5	0.0046	0.8624 0.9997	(54, 380)	(8, 10)	5	0.0056	0.9288 0.9997
	1000	11	(66, 900)	(4, 6)	4	0.0049	0.5357 0.9594	(66, 900)	(4, 6)	4	0.0047	0.5667 0.9596
		25	(87, 830)	(7, 9)	7	0.0050	0.8832 0.9998	(87, 830)	(7, 9)	7	0.0058	0.9646 0.9998
0.0027	50	11	(2, 49)	(4, 6)	4	0.0026	0.3249 0.8627	(2, 49)	(4, 6)	4	0.0026	0.4352 0.9081
		25	(6, 44)	(10, 11)	15	0.0036	0.6179 0.9908	(6, 44)	(10, 11)	15	0.0031	0.7553 0.9947
	100	11	(2, 94)	(3, 5)	3	0.0028	0.3722 0.9040	(2, 94)	(3, 5)	3	0.0025	0.5758 0.9670
		25	(4, 67)	(6, 8)	4	0.0025	0.6595 0.9960	(4, 67)	(6, 8)	4	0.0029	0.8722 0.9987
	500	11	(12, 440)	(3, 5)	4	0.0025	0.4559 0.9479	(12, 440)	(3, 5)	4	0.0031	0.6099 0.9724
		25	(49, 445)	(8, 10)	5	0.0026	0.8250 0.9995	(49, 445)	(8, 10)	5	0.0033	0.9318 0.9995
	1000	11	(55, 900)	(4, 6)	4	0.0027	0.4673 0.9452	(55, 900)	(4, 6)	4	0.0033	0.4673 0.9529
		25	(77, 800)	(7, 9)	7	0.0027	0.8435 0.9997	(77, 800)	(7, 9)	7	0.0032	0.9356 0.9997

Table 6.
Comparison of the AR_{out}*s with the same* FAR *for* Chart 2.

are smaller than the respective ones of *Chart* 1 indicates its efficacy to detect faster the shift of the process from the in-control distribution.

Underlying distributions: exponential with mean equal to 2 (in-control) and 1 (out-of-control), respectively.

It goes without saying that the nonparametric control charts are robust in the sense that their in-control behavior remains the same for all continuous distributions. However, it is of some interest to check over their out-of-control performance for different underlying distributions. We next study the performance of the proposed *Chart* 1 enhanced with the 2-of-3 runs-type rule under normal distribution (θ, δ).

More specifically, the in-control reference sample is drawn from the standard normal distribution, while several combinations of parameters θ, δ have been examined. **Table 5** reveals that the proposed *Chart* 1 with 2-of-3 runs-type rule is superior compared to the existing *Chart* 1 for almost all shifts of the location parameter θ and the scale parameter δ considered.

We next study the out-of-control performance of *Chart* 2 presented in Section 2. In **Table 6**, three different *FAR* levels and several values of the parameters m, n have been considered. For each choice, the *AR* values under two specific Lehmann alternatives corresponding to $\gamma = 0.4$ and $\gamma = 0.2$ are computed via simulation for both *Chart* 2 and *Chart* 2 enhanced with 2-of-2 runs-type rule.

Table 6 clearly indicates that, under a common *FAR*, the proposed *Chart* 2 with the 2-of-2 runs-type rule performs better than *Chart* 2, with respect to *AR* values, in all cases considered. For example, calling for a reference sample of size $m = 500$, test samples of size $n = 11$, and nominal $FAR = 0.0027$, the proposed *Chart* 2 with the 2-of-2 runs-type rule achieves alarm rate of 0.6099 (0.9724) for $\gamma = 0.4$ ($\gamma = 0.2$), while the respective alarm rate for *Chart* 2 is 0.4559 (0.9479).

Tables 7 and **8** shed more light on the out-of-control performance of the proposed *Chart* 2 enhanced with appropriate runs-type rules. More specifically, the schemes being under consideration are designed such as a nominal in-control *ARL* performance is attained. From the numerical comparisons carried out, it is straightforward that *Chart* 2 with 2-of-2 rule becomes substantially more efficient than *Chart* 2. Under Lehmann alternative with parameter $\gamma = 0.5$, the proposed chart exhibits smaller out-of-control *ARL* than *Chart* 2, and therefore it seems more capable in detecting possible shift of the process distribution.

			***Chart* 2**					***Chart* 2 with 2-of-2 runs rule**				
ARL_0	m	n	(a, b)	(j, k)	r	Exact ARL_{in}	ARL_{out}	(a, b)	(j, k)	r	Exact ARL_{in}	ARL_{out}
100	100	5	(4, 99)	(2, 3)	2	115.1	4.7	(10, 80)	(3, 5)	1	116.1	4.6
		11	(5, 78)	(3, 4)	3	88.9	2.8	(10, 83)	(2, 4)	1	83.2	2.7
	500	5	(15, 470)	(2, 3)	2	109.8	5.2	(50, 435)	(2, 3)	1	101.7	4.9
		11	(20, 410)	(4, 6)	4	119.7	6.8	(50, 400)	(2, 3)	1	122.5	2.2
200	100	5	(3, 98)	(2, 3)	2	185.3	6.7	(10, 85)	(2, 5)	1	204.8	1.5
		11	(4, 80)	(3, 4)	3	187.7	3.5	(8, 95)	(2, 3)	1	221	3.3
	500	5	(10, 482)	(2, 3)	2	208.9	7.2	(15, 435)	(3, 5)	1	188.4	5.4
		11	(18, 420)	(4, 6)	4	191.1	7.5	(50, 420)	(2, 3)	1	234.8	2.2

Table 7.
Comparison of the ARL_{out}*s with the same* ARL_{in} *for* Chart 2.

θ	δ	*Chart* 2	*Chart* 2 with 2-of-3 runs rule
0	1	446.6	502.1
0.25	1	163.9	127.9
0.5	1	51.64	26.6
1	1	7.4	3.8
1.5	1	2.1	1.4
2	1	1.2	1.1
0.25	1.25	35.7	35.1
0.5	1.25	17.9	13.7
1	1.25	5.0	3.9
1.5	1.25	2.1	1.7
2	1.25	1.3	1.1
0.25	1.5	15.0	17.7
0.5	1.5	9.8	9.4
1	1.5	4.1	3.6
1.5	1.5	2.1	1.8
2	1.5	1.4	1.3
0.25	1.75	8.5	10.6
0.5	1.75	6.5	7.6
1	1.75	3.5	3.4
1.5	1.75	2.1	1.9
2	1.75	1.5	1.3
0.25	2	5.7	6.9
0.5	2	4.8	5.4
1	2	3.1	3.1
1.5	2	2.0	2.0
2	2	1.5	1.4

Table 8.
Comparison of the $\mathrm{ARL}_{\mathrm{out}}$*s with the same* $\mathrm{ARL}_{\mathrm{in}}$ *for* Chart 2 *under normal distribution* (θ, δ).

In addition, **Table 8** depicts the out-of-control *ARL* performance of *Chart* 2 under normal distribution. More precisely, several shifts of both location and scale parameter have been considered, and *Chart* 2 with 2-of-3 rule detects the underlying shift sooner than *Chart* 2 does in almost all cases examined.

Finally, **Tables 9** and **10** present simulation-based comparisons of the nonparametric monitoring scheme *Chart* 3 with 2-of-2 runs rule versus *Chart* 3 established by Triantafyllou [7]. For delivering the numerical results displayed in **Tables 9** and **10**, the Lehmann alternatives have been considered as the out-of-control distribution. Each cell contains the *AR*s attained for $\gamma = 0.5$ (upper entry) and $\gamma = 0.2$ (lower entry).

5. Conclusions

In the present chapter, we investigate the in- and out-of-control performance of distribution-free control charts based on order statistics. Several runs-type rules are

			Chart 3					*Chart* 3 with 2-of-2 runs rule				
FAR	*m*	*n*	*(a, b)*	*(i, c, j, d)*	*r*	*Exact FAR*	AR_{out}	*(a, b)*	*(i, c, j, d)*	*r*	*Exact FAR*	AR_{out}
0.01	100	11	(6, 93)	(7, 2, 4, 7)	4	0.0104	0.5363 0.9537	(15, 75)	(7, 2, 4, 7)	1	0.0086	0.7064 0.9825
		25	(5, 86)	(6, 2, 6, 12)	10	0.0102	0.8181 0.9992	(10, 79)	(6, 2, 6, 12)	1	0.0108	0.9103 0.9992
	500	11	(31, 484)	(33, 2, 4, 7)	6	0.0108	0.5335 0.9583	(80, 300)	(33, 2, 4, 7)	1	0.0088	0.7326 0.9876
		25	(30, 450)	(49, 2, 7, 9)	8	0.0101	0.9122 0.9999	(75, 300)	(49, 2, 7, 9)	1	0.0089	0.9585 0.9999
	1000	11	(60, 967)	(65, 2, 4, 7)	6	0.0099	0.5311 0.9584	(155, 690)	(65, 2, 4, 7)	1	0.0096	0.7161 0.9873
		25	(30, 960)	(99, 2, 7, 9)	8	0.0098	0.9181 0.9999	(165, 600)	(99, 2, 7, 9)	1	0.0129	0.9778 0.9999
0.005	100	11	(4, 93)	(5, 2, 4, 7)	4	0.0051	0.4134 0.9198	(14, 75)	(5, 2, 4, 7)	1	0.0054	0.6312 0.9787
		25	(4, 88)	(5, 2, 6, 12)	10	0.0051	0.7508 0.9983	(10, 81)	(5, 2, 6, 12)	1	0.0051	0.9227 0.9989
	500	11	(26, 478)	(31, 2, 4, 7)	5	0.0052	0.5094 0.9536	(70, 300)	(31, 2, 4, 7)	1	0.0049	0.6653 0.9827
		25	(30, 450)	(43, 2, 7, 9)	8	0.0052	0.8755 0.9998	(70, 314)	(43, 2, 7, 9)	1	0.0046	0.9485 0.9998
	1000	11	(60, 985)	(65, 2, 4, 7)	6	0.0050	0.5299 0.9583	(150, 700)	(65, 2, 4, 7)	1	0.0047	0.7051 0.9870
		25	(30, 960)	(87, 2, 7, 9)	8	0.0050	0.8832 0.9998	(150, 600)	(87, 2, 7, 9)	1	0.0056	0.9645 0.9998
0.0027	100	11	(3, 94)	(4, 2, 4, 7)	4	0.0029	0.3410 0.8905	(11, 75)	(4, 2, 4, 7)	1	0.0032	0.5283 0.9660
		25	(5, 86)	(6, 2, 7, 9)	7	0.0025	0.6951 0.9971	(9, 80)	(6, 2, 6, 9)	1	0.0024	0.8803 0.9988
	500	11	(24, 480)	(28, 2, 4, 7)	4	0.0028	0.4715 0.9453	(70, 345)	(28, 2, 4, 7)	1	0.0024	0.6624 0.9834
		25	(30, 450)	(38, 2, 7, 9)	8	0.0027	0.8338 0.9996	(70, 320)	(38, 2, 7, 4)	1	0.0030	0.9497 0.9996
	1000	11	(53, 990)	(55, 2, 4, 7)	6	0.0027	0.4673 0.9452	(135, 705)	(65, 2, 4, 7)	1	0.0023	0.6445 0.9823
		25	(30, 960)	(78, 2, 7, 9)	8	0.0028	0.8479 0.9997	(145, 600)	(78, 2, 7, 9)	1	0.0029	0.9587 0.9997

Table 9.
Comparison of the AR_{out}*s with the same* FAR *for* Chart 3.

employed in order to enhance the ability of the aforementioned nonparametric monitoring schemes to detect possible shifts in distribution process. The *AR* and the *ARL* behavior of the underlying control charts is studied under several out-of-control situations, such as the so-called Lehmann alternatives and the exponential or the normal distribution model. The numerical experimentation carried out depicts the melioration of the proposed schemes with the runs-type rules. It is of some research interest to branch out the incorporation of such runs rules (or even more complicated) to additional nonparametric control charts based on well-known test statistics.

			Chart 3					Chart 3 with 2-of-2 runs rule				
ARL_0	m	n	(a, b)	(i, c, j, d)	r	Exact ARL_{in}	ARL_{out}	(a, b)	(i, c, j, d)	r	Exact ARL_{in}	ARL_{out}
100	50	10	(1, 45)	(2, 1, 2, 4)	4	101.06	2.3 1.02	(3, 40)	(2, 1, 2, 4)	1	111.4	2.3 1.02
	100	5	(3, 100)	(4, 1, 2, 4)	3	123.51	4.73 1.22	(13, 78)	(4, 1, 2, 4)	1	102.4	4.72 1.14
	100	10	(2, 89)	(3, 1, 2, 4)	3	101.82	2.31 1.02	(12, 81)	(5, 5, 3, 5)	1	114.0	2.28 1.02
200	50	10	(1, 48)	(3, 1, 2, 4)	3	213.49	2.30 1.02	(3, 45)	(2, 1, 2, 4)	1	187.5	2.23 1.02
	100	5	(1, 100)	(3, 1, 2, 4)	2	244.88	6.59 1.29	(10, 80)	(3, 1, 2, 4)	1	230.9	5.62 1.24
	100	10	(4, 94)	(5, 5, 3, 5)	4	222.69	3.20 1.04	(10, 82)	(5, 5, 3, 5)	1	192.2	3.10 1.04

Each cell contains the AR's attained for Y = 0.5 (upper entry) and Y = 0.2 (lower entry).

Table 10.
Comparison of the ARL_{out}*s with the same* ARL_{in} *for* Chart *3.*

Author details

Ioannis S. Triantafyllou
Department of Computer Science and Biomedical Informatics, University of Thessaly, Greece

*Address all correspondence to: itriantafyllou@uth.gr

References

[1] Shewhart WA. Quality control charts. Bell System Technical Journal. 1926;**2**:593-603

[2] Montgomery DC. Introduction to Statistical Quality Control. 6th ed. New York: John Wiley; 2009

[3] Qiu P. Introduction to Statistical Process Control. New York: CRC Press, Taylor & Francis Group; 2014

[4] Chakraborti S, van der Laan P, van der Wiel MA. A class of distribution-free control charts. Journal of the Royal Statistical Society: Series C: Applied Statistics. 2004;**53**:443-462

[5] Balakrishnan N, Triantafyllou IS, Koutras MV. A distribution-free control chart based on order statistics. Communications in Statistics: Theory and Methods. 2010;**39**:3652-3677

[6] Triantafyllou IS. Nonparametric control charts based on order statistics: Some advances. Communications in Statistics: Simulation and Computation. 2018;**47**:2684-2702

[7] Triantafyllou IS. A new distribution-free monitoring scheme based on order statistics. Journal of Nonparametric Statistics. 2019;**31**:1-30

[8] Koutras MV, Triantafyllou IS, editors. Recent advances on univariate Shewhart type distribution free control charts. In: Distribution-free Methods for Statistical Process Monitoring and Control. Springer; 2019. DOI 10.1007/978-3-030-25081-2

[9] Chakraborti S, Graham M. Nonparametric Statistical Process Control. USA: John Wiley & Sons; 2019

[10] Qiu P. Some perspectives on nonparametric statistical process control. Journal of Quality Technology. 2018;**50**:49-65

[11] Qiu P. Some recent studies in statistical process control. In: Statistical Quality Technologies. 2019. pp. 3-19

[12] Klein M. Two alternatives to the Shewhart X control chart. Journal of Quality Technology. 2000;**32**:427-431

[13] van der Laan P, Chakraborti S. Precedence tests and Lehmann alternatives. Statistical Papers. 2001;**42**:301-312

Chapter 13

Combining Capability Indices and Control Charts in the Process and Analytical Method Control Strategy

Alexis Oliva and Matías Llabrés

Abstract

Different control charts in combination with the process capability indices, C_p, C_{pm} and C_{pk}, as part of the control strategy, were evaluated, since both are key elements in determining whether the method or process is reliable for its purpose. All these aspects were analyzed using real data from unitary processes and analytical methods. The traditional x-chart and moving range chart confirmed both analytical method and process are in control and stable and therefore, the process capability indices can be computed. We applied different criteria to establish the specification limits (i.e., analyst/customer requirements) for fixed method or process performance (i.e., process or method requirements). The unitary process does not satisfy the minimum capability requirements for C_p and C_{pk} indices when the specification limit and control limits are equal in breath. Therefore, the process needs to be revised; especially, a greater control in the process variation is necessary. For the analytical method, the C_{pm} and C_{pk} indices were computed. The obtained results were similar in both cases. For example, if the specification limits are set at ±3% of the target value, the method is considered "satisfactory" ($1.22 < C_{pm} < 1.50$) and no further stringent precision control is required.

Keywords: control chart, average run length, capability indices, critical values, lower confidence bound

1. Introduction

At first, it is impossible to determine the quality of a product. However, it is necessary that the manufacturing processes are in control and stable as well as all the involved unitary processes in order to reduce the process variability. When an analytical procedure is performed on a sample, this is itself a process just as the manufacturing operation is a procedure. By analogy, we can apply the same rules and principles.

Control charts, as online statistical process control procedure, represent the first option for achieving this objective. Statistical process control allows to analyze the process stability and to estimate the process capability, knowing the level of variability. The Shewhart control chart is the most used technique to detect statistical changes in process quality. Walter A. Shewhart of the Bell Telephone Laboratories developed it in 1924. The control chart can be used as an estimating device, for example, process parameters such as the mean, standard deviation, fraction nonconforming, and so on may be estimated from a control chart. In addition, these estimates may be used to determine the capability of the process to produce acceptable results [1]. Shewhart control charts are effective when the in-control process data are stationary (i.e., the process data vary around a fixed mean in a stable manner) and uncorrelated. Under these conditions, their performance is predictable, allowing out-of-control situations to be reliably detected. In this type of control chart, the first step is as follows: a set of process data are collected and analyzed all at once in a retrospective analysis, constructing different control limits (such as warning and action control limits) in order to verify if the process is in control over the time during the collection of data. Second is to check if these limits can help to monitor future productions or samples. Alternatively, chart based on standard values allows specifying standard values for the process mean and standard deviation without analysis of the past data. A limitation of Shewhart control charts is that it uses only the information about the process contained in the last analyzed sample, ignoring any information provided by the set of collected data. This fact makes the Shewhart control chart relatively insensitive to small process shifts, about 1.5 standard deviations or less. The exponentially weighted moving average (EWMA) and the cumulative sum (Cusum) control charts are two good options in those situations where it is important to control small process shifts. Roberts [2] and later Crowder [3] and Lucas and Saccucci [4] introduced the EWMA control chart which analyzed different aspects of interest in detecting small changes in the process. Other authors, such as Lucas [5], Hawkins and Olwell [6], indicate that the Cusum control is more effective than the traditional Shewhart control chart in this type of situations.

In industrial activities or in the laboratory, it is necessary to obtain information about the performance of the process or analytical method when it is operating under statistical control. For this, the process or analytical method is in control and stable. Process capability indices (PCIs) give an indication of the capability of a process or analytical method [7]. They are designed to quantify the relation between the desired specifications and the actual performance of the process or analytical method. In addition, the capability indices are calculated, to evaluate whether the process under study is able to provide sufficient conforming units. The capability indices could be used to evaluate whether the analytical method is only able to provide enough conforming results to check if a method is fitted for its intended purpose [8]. Various examples of the usefulness of capability indices in the framework of analytical method validation can be found in the literature [9–11]. At first, the methodology described earlier can be applied to any process or analytical method in statistical control.

The main objective of this work was to evaluate the use of control charts in combination with the process capability indices as key elements in determining whether the process or analytical method is fitted for its purpose. The process capability indices, C_p, C_{pk}, and C_{pm}, were computed. The level of variability (i.e., method or process performance) was evaluated through the control chart, whereas the method or process specifications (i.e., analyst/customer requirements) were analyzed under different criteria based on the specification limit range. Finally, to determine whether the process or method meets the present capability requirement and runs under the desired quality conditions, the Pearn and Shu [24] method was used. All these aspects were analyzed using two examples: (1) the upper punch compaction force data obtained in a tablet manufacture process and (2) the RP-HPLC method data used for insulin quantitation in pharmaceutical preparations [12].

2. Control chart in unitary process

2.1 X-bar and MR-control chart

The stability of a process is an important property, since if it is stable in the current time frame, it is also likely to be so in the future, assuming that no major changes occur [13]. This means that the process variation is due only to random causes and all assignable or special causes have been removed. If this is fulfilled, one can draw conclusions about the process capability and use the result for predicting it in future or other conditions. Usually, the process mean is monitored using location charts such as the x-chart, and the process dispersion is monitored using dispersion charts such as the R- or S-chart [1]. These control charts are based on samples (or subgroups) of n observations taken at regular sampling intervals. There are, however, many applications in which the control charts are based on individual observations ($n = 1$) rather than samples of $n > 1$. In such cases the R-chart cannot be used, as it is impossible to calculate the within-sample variation when the sample size equals 1.

The control charts discussed above are designed under the assumption that a process being monitored will produce measurements that are independent and identically distributed over time, when only the inherent sources of variability are present in the process. For this, it is necessary to check the normality of the data, which is assessed through Q-Q plots and using statistical tests (e.g., Anderson-Darling, Shapiro-Wilk, or chi-square). **Figure 1** shows the Q-Q plots for tablet manufacturing process. Shapiro-Wilk test confirmed that data follow a normal distribution at 5% significance level.

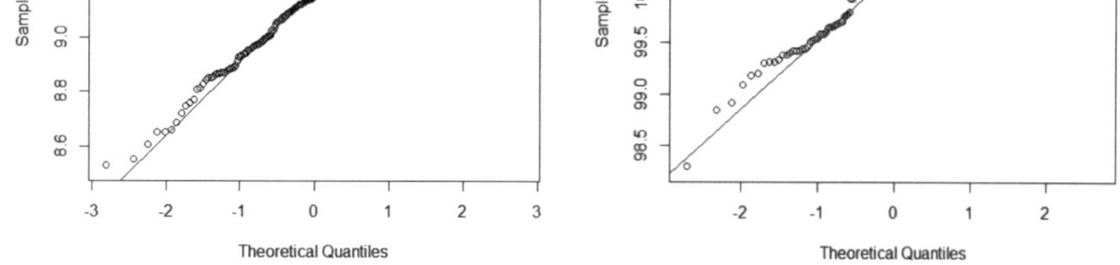

Figure 1.
Normal Q-Q probability plot for compaction process data (left) and HPLC analytical method data used for the insulin quantification (right).

Figure 2 shows the two control charts, one for monitoring the process center (x-bar chart) and the other for monitoring the process variation (MR-chart), when a separate observation is made at each sampling point [1].

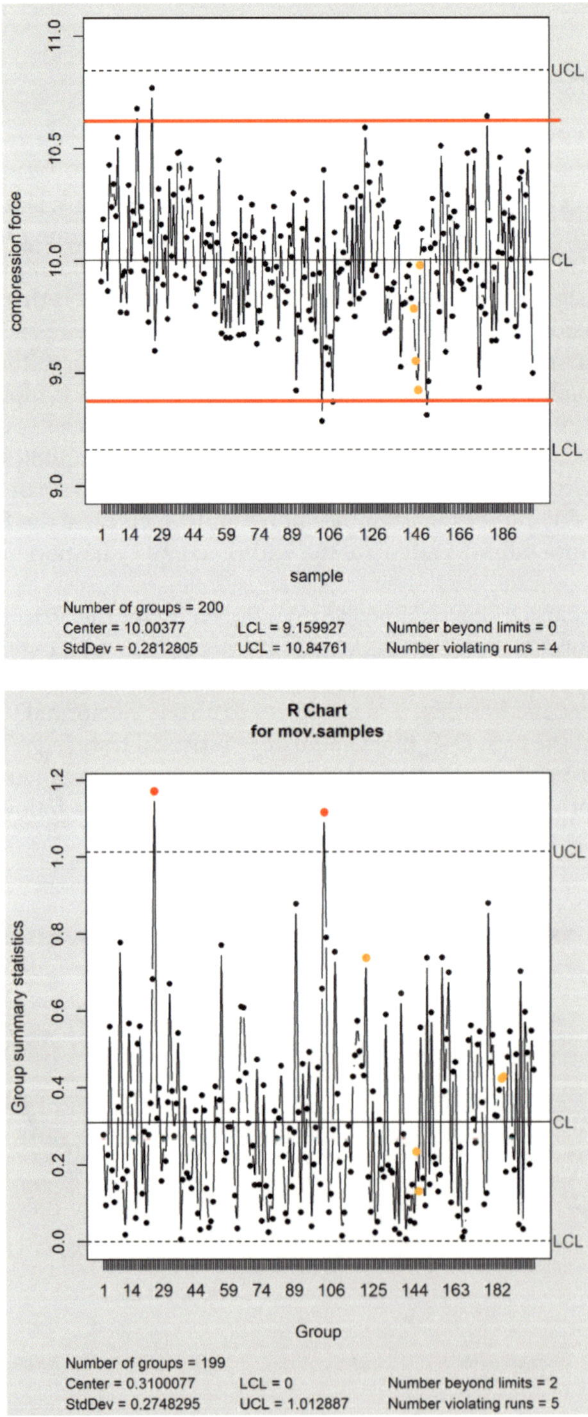

Figure 2.
Control charts for compaction process: (upper) x-bar control chart and (lower) MR-chart (UCL = upper control limit; LCL = lower control limit; CL = mean or average range for the MR-chart). The red line corresponds to the warning limits using a RSD of 6%.

For the x-bar control chart based on individual observations, the central lines (CL) and control limits (UCL and LCL) are:

$$UCL = \overline{X} + 3\frac{\overline{MR_2}}{d_2}; CL = \overline{X}; \mathrm{LCL} = \overline{X} - 3\frac{\overline{MR_2}}{d_2} \tag{1}$$

where x-bar is the sample mean and MR2 is the mean moving range of length two.

In this case, the traditional choice is to use the moving range chart (MR-chart), which is the range of successive individual observations, to detect changes in the process variation [1].

$$MR_i = |x_i - x_{i-1}| \tag{2}$$

For the moving range charts, the following equations with n = 2 are used to establish the CL and control limits, respectively:

$$LCL = \overline{MR_2} \cdot D_4; CL = \overline{MR_2}; UCL = \overline{MR_2} \cdot D_4 \tag{3}$$

In formulating the control limits of x-bar and MR-control charts, several factors, d_2, D_3, and D_4, are constant, dependent on n, and assuming normal data distribution [14]. These values are tabulated and can be found in the bibliography [1]. In our case, d_2 = 1.128 and D_4 = 3.268, respectively.

In this first example, we used the upper punch compaction force as variable. The data used in this example were generated using a compress machine (model Korsch AG XP-1) for 10 min with a sample frequency of 20 samples/min. The collected data corresponds to the acetaminophen tablet batch to laboratory scale (data not published).

The data analysis was performed using the R-program (version 3.6.1) and plotted using the "qcc" package [15].

Given the approximate normality of the data, we can use the x-chart to estimate the process mean, obtaining a value of 10.0 kN, whereas the MR-chart provides the process standard deviation, obtaining a value of 0.31 kN.

Figure 2 shows the x-bar control chart for the compression force variable. All plotted values fall within the control limits (9.16, 10.85), and therefore, the process is in statistical control. In addition, there is no evidence of cyclical or periodic behavior. However, there is a located zone between samples 148 and 152 indicating the nonrandom patterns present. This situation is related with the presence of "eight consecutive points plot on one side of the center line" [1] according to the application of decision rules for detecting this type of variation.

In the compaction process, it is usually to fix the warning limits at ±6% of the mean value (RSD = 6%). In such situation, there are six points beyond these limits, indicating the existence of a problem during the process. The MR-control charts exhibit two points above the upper control limits (UCL = 1.013), and therefore the process should be considered out of control (**Figure 2**). However, a point above the upper control limit followed immediately by a point below control limit would not signal an out-of-control alarm. A similar situation was observed when the warning limits were fixed at ±6% of the mean value. In addition, the control charts show other forms of nonrandom variation; all of them are due to the presence of "eight consecutive values on one side of the centerline." It is true that, when a point is plotted outside of the action limits, a search for an assignable cause is made and corrective action is taken if necessary. We have no explanation for this. The causes could be various: particle size and shape distribution, flow properties of the bulk material, mix process, tablet weight, etc. In this last case, we weighted some tablet during the manufacture process, the mean value being 700 ± 5 mg (n = 40), but this tablet batch does not satisfy the proposed fragility test by the USP [16]. Therefore, the process may not be operating properly. In this case, the sensitivity of the control chart should improve, changing the sampling frequency and/or the sample size in order to obtain more information about the process.

This strategy may increase the risk of false alarms and be confusing to the operating personnel. In such situation, the average rung length (ARL) of the control chart is a good alternative. The ARL is the average number of points that must be plotted before a point indicates an out-of-control condition. If the process observations are uncorrelated, then in any Shewhart control chart, the ARL can be calculated as:

$$ARL = \frac{1}{p} \tag{4}$$

where p is the probability that any point exceeds the control limits. For the control at three sigma limits, p = 0.0027, and therefore, an out-of-control signal will be generated every 370 samples, on the average, even if the process remains in control. Some analysts like to report percentiles of the run length distribution instead of just the ARL [1]. The 10th and 50th percentiles are used more. In our example, around 10% of the time, the in-control run length will be less than 38 samples, and 50% of the time, it will be less than 256 samples.

For the MR-control chart, the probability that any point exceeds the upper control limit (1.013) is 0.0052, and therefore, ARL is 190; this supposes that an out-of-control sample will be generated every 190 samples on the average. The second point outsider of control limits (sample #98) is too far from this value (see **Figure 2**).

2.2 Cusum and EWMA control chart

Cumulative sum and exponentially weighted moving average control charts efficiently complement the x-bar and MR-control charts when there is interest in detecting small changes in the process, around ±1.5 SD, and the sample consists of an individual unit. However, many researchers have discussed which of them is better in accordance with the level of variability that must be detected [17]. In practice, the EWMA control chart worked well with the parameters λ = 0.4 (smoothing constant) and L = 3.054 (control limit width fixed at three standard deviations), a value recommended by Montgomery [1]. Under this scenario, the sample number 112 was out of the control limits (see **Figure 3**), whereas the sample number 153 is very close to this limit. Using a value of λ = 0.2, the change was clearer, since the samples closest were also affected (data not shown). The values of λ = 0.2 and 0.05 with the respective width of control limits L = 2.814 and 2.614 are the best option to detect the average changes at order of one standard deviation.

Cusum control charts directly incorporate all the information into the sequence of sample values by plotting the cumulative sums of their deviations from a value objective [18]. Moreover, when a tendency up or down appears, it indicates the process average changes, which requires a search to determine the causes. Oliva and Llabrés described a similar situation [19]. The Cusum control chart showed a similar situation to those observed in the EWMA control chart for the sample number 112; it was out of the limits. When we fixed the shift detection at ±1 SD, the situation is totally different; 17 points were beyond boundaries, approximately 8.4%, especially located at the beginning and at the end of the process (data not shown); for a shift detection equal to ±1.5 SD, a unique value was out of this limit. However, the data

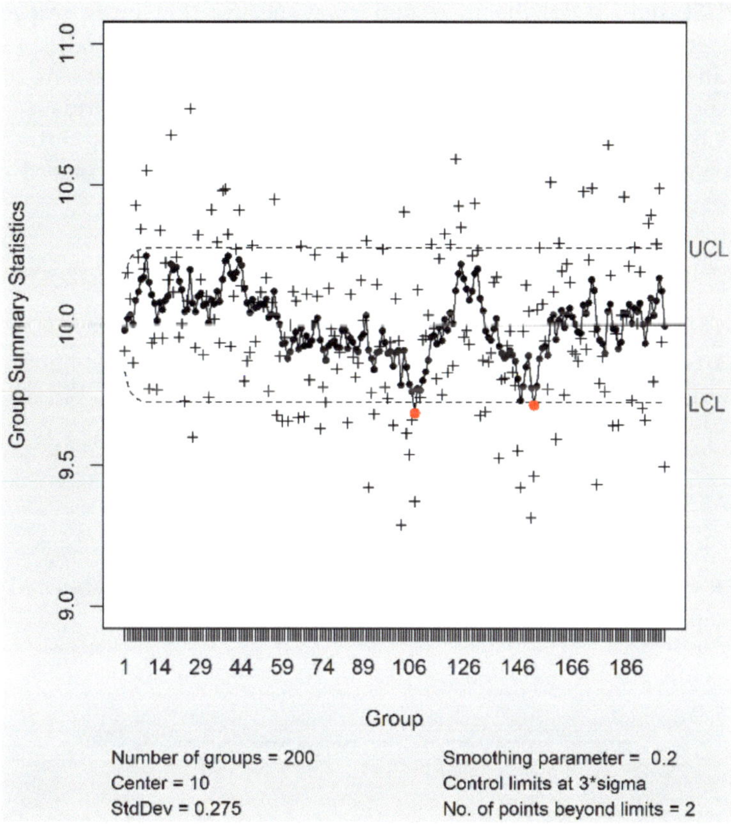

Figure 3.
EWMA control chart for the compaction process with the parameters λ = 0.2 and L = 3.054. Under these conditions, two points were beyond limits (#110 and #153, in red), whereas the number of points beyond limits was zero for λ = 0.4.

seems to describe a random way with an average of zero. There is no zone with an upward (C+) or downward (C−) tendency, perfectly defined, typical behavior of the Cusum control charts when the average process changes.

At first, the Cusum control chart performed better for detecting shifts lower than 1.5 SD. However, EWMA provided the forecast of where the average will be in the next period, which makes it easier to apply in the process control (**Figure 4**).

2.3 Computing the process capability indices and the specification limits

Details about PCIs and their statistical properties can be found in the literature [20, 21]. A capability index is generally a function of the process parameters, such as the mean μ, standard deviation σ, target value T, lower specification limit (LSL), and upper specification limit (USL) of x variable.

The C_{pm} index is the best option to drive the process (or method) to the target value since this is intended to account for variability from the process (or method) mean and deviation from the target value T [7]. For a normally distributed process that is demonstrably stable (under statistical control), Boyles [22] considered the maximum likelihood estimator of C_{pm} as:

$$C_{pm} = \frac{USL - LSL}{6\sqrt{\sigma^2 + (\mu - T)^2}} \quad (5)$$

where USL and LSL are the upper and lower specification limits, respectively. Their difference provides a measure of allowable process (or method) spread (i.e., customer/analyst requirements), whereas σ^2 and $(\mu - T)^2$ are a measure of precision and accuracy, respectively (i.e., process or method performance requirements). The mean of the process (or method) μ is estimated through the sample mean x-bar, whereas the following estimator for the standard deviation σ can be used:

$$\hat{\sigma} = \sqrt{\frac{\sum s_i^2}{m}} \tag{6}$$

where s_j is the standard deviation of each subgroup and m is the number of subgroups. If the process is monitored using the MR-control chart, the following estimator can be used:

$$\hat{\sigma} = \frac{\overline{R}}{d_2} \tag{7}$$

where $\overline{R}$ and d_2 are the average range (in our case, MR2) and tabulated constant that only depend on the sample size n, respectively [1].

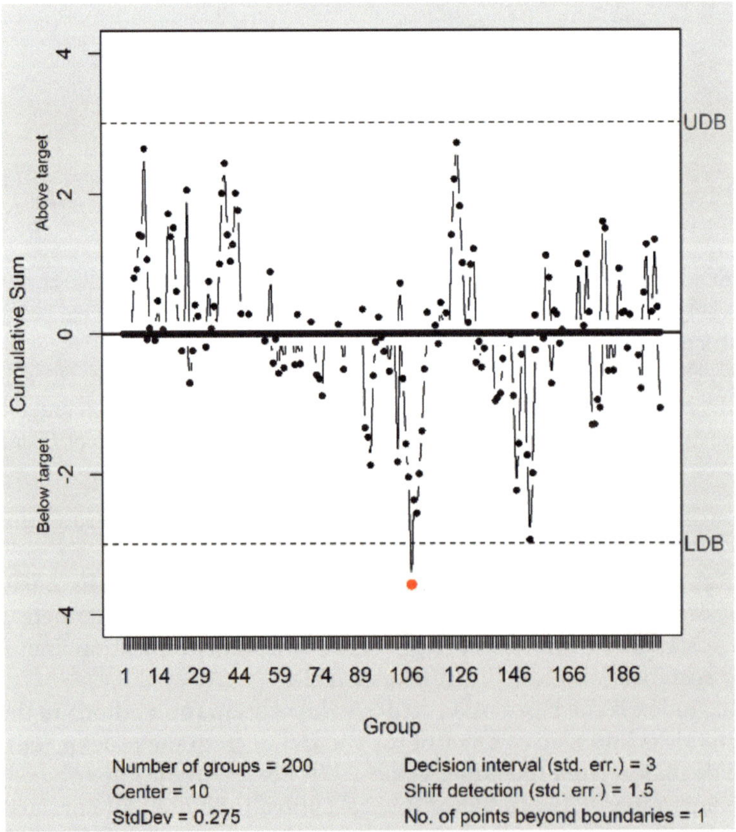

Figure 4.
Cusum control chart for the compaction process. The number of points beyond boundaries was one (#110, in red) for shift detection fixed at ±1.5 SD, whereas the number of points was 8.4% for detection fixed at ±1 SD.

The C_{pk} index is defined as the ratio of the minimal distance of the specification limits to the method average to three times the standard deviation of the method (if the average is in between the specification limits) [23].

$$C_{pk} = \min\left\{\frac{USL-\mu}{3\sigma}, \frac{\mu-LSL}{3\sigma}\right\} \tag{8}$$

C_{pk} is more commonly used because it is not dependent on the process or method being centered. However, C_{pm} is more sensitive to departure from the method target than C_{pk} is [24]. For example, when μ is within the interval of the specification limits, C_{pk} depends inversely on the method standard deviation σ (i.e., systematic error, σ^2) and becomes large as σ gets closer to zero. C_{pk} also depends on the distance of the mean from the specification limits (i.e., method centering).

If the method precision is improved, the C_{pm} will increase. If the method drifts from its target value (i.e., if μ moves away from T), then C_{pm} decreases. When both the method precision and the mean are modified, the C_{pm} index reflects these changes as well. This is also true for the C_{pk} index.

Pearn and Shu [24] proposed the lower confidence bounds "C" on C_{pm} to measure the minimum capability of the process, based on the sample data. In this case, the critical values (Co) are used for making decisions in method capability testing with a designated type-I error, α, which is the risk of misjudging an incapable method (Ho: $C_{pm} \leq C$) as a capable one (H1: $C_{pm} > C$), where C is the required process capability. This supposes that the decision-making procedure ensures that the risk of making a wrong decision will be no greater than the preset type-I error α. The algorithm proposed by Pearn and Shu [24] was used to compute the lower confidence bounds C. For this, the sample of size n, the confidence level γ (0.95), the estimated value C_{pm}, and the parameter ξ must be provided. In practice, the parameter ξ = (μ − T)/σ is unknown, but it can be calculated from the sample data as $\xi = \left(\overline{X} - T\right)/S_n$, S_n being the process standard deviation.

Pearn and Chen [25] and Pearn et al. [26] have developed a procedure to obtain the lower confidence bounds and critical values of C_p and C_{pk} to determine whether a process or method meets the capability requirement or not.

To calculate the PCIs, it is necessary to know the inherent variability in a process (using the control chart) and the customer requirements in terms of specification limits [27]. Control limits are set by the process and formulas; they are the voice of the process. The specification limits (LSL, USL) may be flexible, not rigorous, based on different criteria, since they represent the voice of the customer [7, 28]. The focus is to set some specification limits and compare them with the control limits of the process since they are the voice of the performance of the process (**Figure 5**).

Bouabidi et al. [8] proposed fixing the specification limits at ±5% around the true or nominal value, although Oliva and Llabrés [29] have proposed a lower variation level. The true value can be calculated using different procedures depending on variable characteristics. Other criteria could be to fix the specification limits equal to the control limits, which are just μ ± 3σ if a normal distribution is assumed.

Since the method is in control, capability indices can be computed, in this case, the indices C_p and C_{pk} (**Table 1**). To calculate the C_{pm}, the method mean and variability must be estimated relative to the method target and specification limits [25]. In this case, the T value is unknown given the process characteristic; no independent approach is available to calculate it since this response depends on working conditions. If the fixed T value is equal to the process mean, this implies that C_{pm} is equal to the C_p index.

$$C_p = \frac{USL - LSL}{6\sigma} \tag{9}$$

With respect to specification limits, we cannot apply Bouabidi et al.'s [8] proposed criteria, based on variations around the target value T. Other criteria could be to fix the specification limits equal to the control limits. In this case, the C_{pk} index was 1.03. To determine if the process meets the capability requirement, we must calculate the critical value Co for C_{pk} based on α risk, sample size, and C value (i.e., the required process capability) [25]. We find the critical value Co =1.095, based on C = 1.0, α = 0.05, and sample size n = 200, demonstrating that the process fails to meet the capability requirements (**Table 1**).

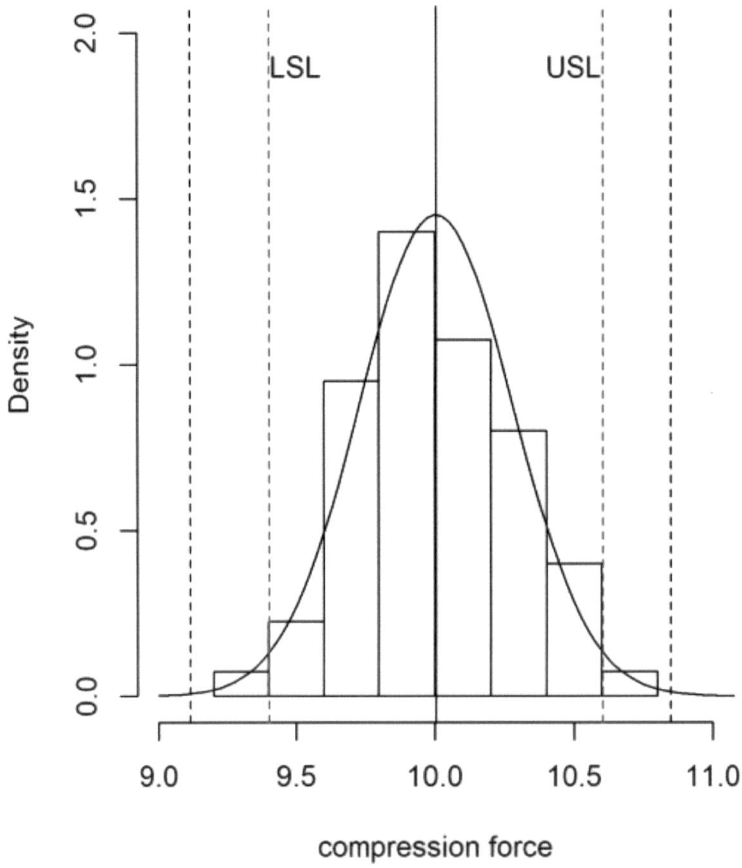

Figure 5.
The black-dashed line shows the specified limits (USL and LSL) established at ±10% of mean value, whereas the red-dashed line corresponds to limits at ±6% of the mean value. The black line is the process mean.

USL-LSL	C_p	Co	C_{pk}	Co	Process capability
±3SD	1.02	1.081	1.03	1.095	Inadequate
±6% of x-bar	0.73	1.081	0.73	1.095	Inadequate
±10% of x-bar	1.21	1.081	1.21	1.095	Capable

Table 1.
C_p and C_{pk} values as a function of the specification limit (USL-LSL). The process capability is based on the critical values (Co) according to Pearn et al. [26].

A similar result was obtained for the C_p index. A value of 1.02 was obtained, whereas the critical value Co with α risk of 0.05 was 1.081 [26], and therefore, the process does not satisfy the minimum process capability requirements.

An alternative way to increase the process capability is to improve the process performance (modifying the allowable process spread through specification limits) or reduce the process systematic error (i.e., process standard deviation). In this last case, the quality improvement effort should focus on reducing the process variation, for example, changing the sampling frequency could solve the problem.

The process performance may modify the function of the specification limit width. In the compaction process, it is usually to fix the warning limits at ±6% of the mean value (RSD = 6%). If the specification limits are fixed at this level, the estimated C_{pk} is lower than the critical value ($0.73 < 1.095$), and therefore, the process is not capable. If the specification limits are fixed at ±10% of the process mean value, the estimated C_{pk} value is 1.23 and exceeds the critical value of 1.095, indicating that the process is capable. This option does not suppose a real improvement in the process capability since the process conditions are maintained. The control limits are driven by the natural variability of the process, whereas the specification limits are determined externally by the manufacturing engineering, the customer, etc. We should know the process variability when setting specification limits. In our opinion, it is necessary to establish a compromise between the specification limit width and process variability.

3. Control chart in analytical method

3.1 X-bar and MR-control chart

The main objective of any validation process is to check the maintenance of validation conditions in the laboratory over a long time period. In this second example, we used the insulin peak area expressed as concentration (U/mL) as control parameters [12]. For this, a standard solution with a nominal concentration of 100 U/mL was analyzed each working day (n = 144). The predicted concentration for the standard solution was obtained from the method calibration. This value is not independent due to the measurement errors which depend on various factors related to the method and its validation but not on the analyst [29].

Histogram and normal probability plots show that the collected data follow the normal distribution (**Figure 1**). The Shapiro-Wilk test confirmed this assumption. Therefore, control charts can be used to obtain the method requirements.

The method mean was estimated to be 100.227 U/mL from the x-bar control chart (**Figure 6**), while the method standard deviation was estimated to be 0.60 U/mL from the MR-chart. The control limits were estimated using the "qcc" package from the R-program [15].

The x-bar control chart shows that all plotted values fall within the control limits (98.40, 102.06), and therefore, the method is in statistical control. In addition, there is no evidence of cyclical or periodic behavior. However, the sample (#74) was outside of the control limits, but the cause of this was attributable to introducing a new column, whereas the sample #86 was related with the presence of "eight consecutive points plot on one side of the center line" [1]. The application of decision rules for detecting nonrandom patterns on control charts indicates that, in this situation, the method is out of control. However, the use of these rules allows enhancing the sensitivity of control charts against only criterion of control limit violation.

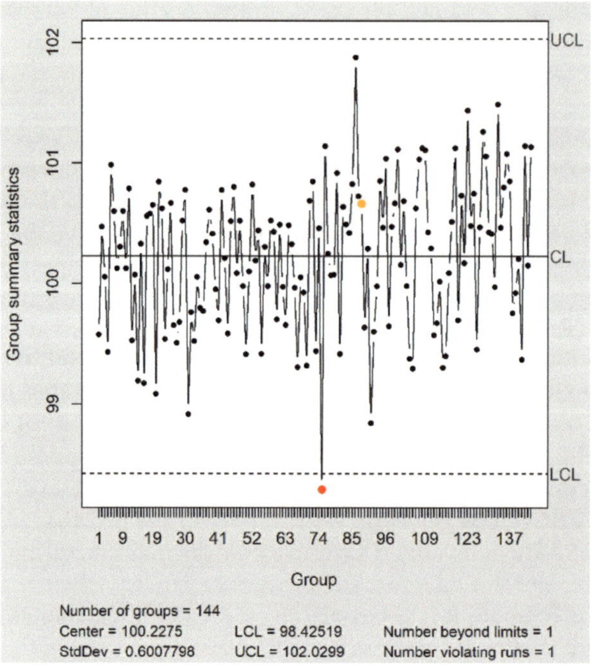

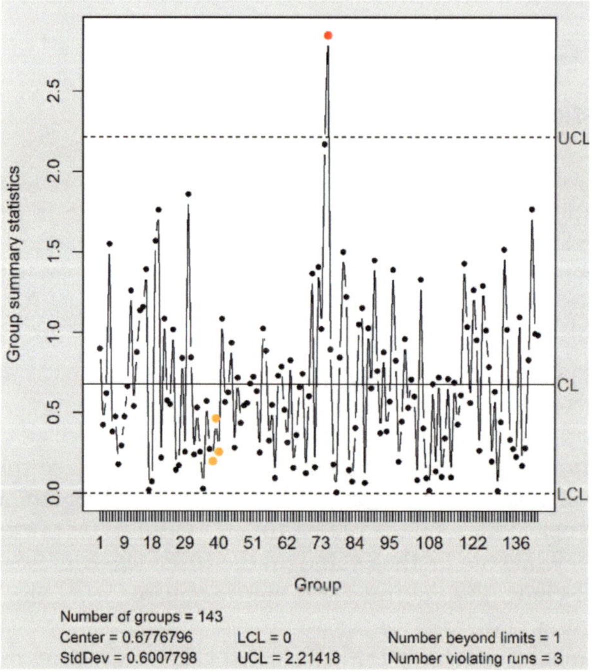

Figure 6.
Control charts for HPLC method used for the insulin quantification in pharmaceutical preparations: (upper) x-bar control chart and (lower) MR-chart. (UCL = upper control limit; LCL = lower control limit; CL = mean or average range for the MR-chart).

The ARL was 370 since the probability that any point exceeds the control limits is 0.0027.

The MR-control charts exhibit one point above the upper control limits (UCL = 2.214) as well as other forms of nonrandom variation, around the sample #40, and therefore the method should be considered out of control (**Figure 6**).

In such situation, it is necessary to search the cause and take corrective action. In the first case, the cause was assignable with a column change, whereas the second one was due to the presence of "more eight consecutive point plots on one side of the centerline."

In this case, the obtained ARL value for MR-control was 189 since the probability that any value exceeds the upper control limit was 0.0053.

3.2 Cusum and EWMA control chart

The data were analyzed using the Cusum and EWMA control charts.

The EWMA control chart shows that all plotted values fall within the control limits (**Figure 7**) using a smoothing constant of 0.2 (λ = 0.2) and control limit width fixed at three standard deviations (L = 3.054).

Cusum control chart, with a shift detection fixed at ±1 SD, shows four points beyond boundaries, all of them greater than the upper control limit. The first alteration is located around the samples #84–85, whereas the second one appears close to the end of the process (#130–131). In addition, if both alterations presented an upward tendency, it indicates the process average changes, which requires a search to determine the causes. When we fixed the shift detection at ±1.5 SD, the situation is totally different, all points fall within control limits (data not shown), and the data describe a random way with an average of zero, since the points show no evidence of an upward or downward tendency (**Figure 8**).

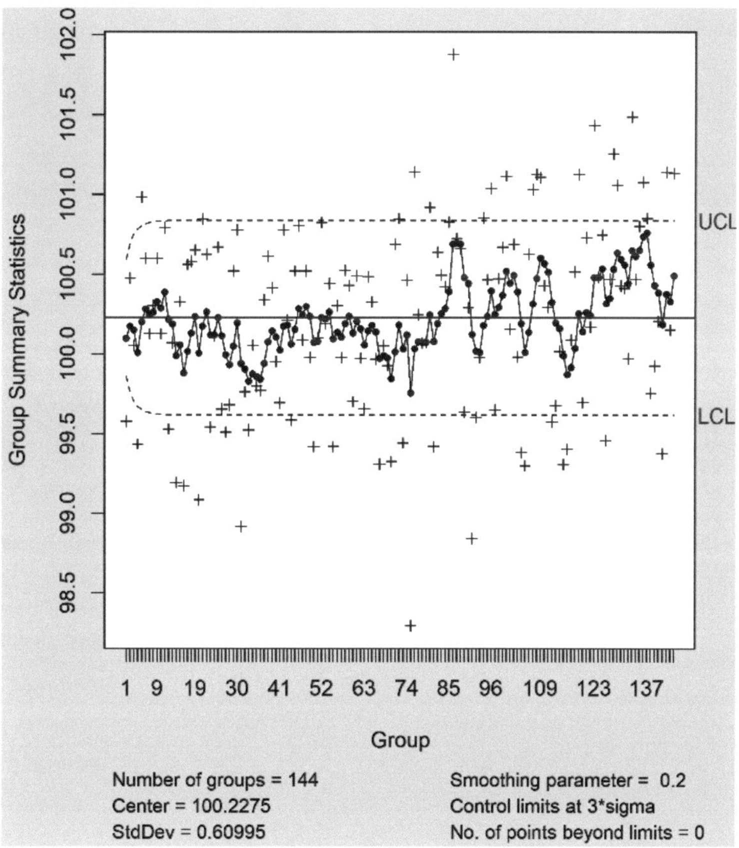

Figure 7.
EWMA control chart for HPLC method. All points fall within control limit for λ = 0.2 and L = 3.054.

3.3 Process capability and specification limits

Since the analytical method is in control and stable, capability indices can be computed. **Table 2** shows the estimated C_{pk} and C_{pm} indices to analyze the capability of our analytical method. The index C_{pm}, sometimes called the Taguchi index, adequately reveals the ability of the method to cluster around the target. This reflects the degree of method targeting (centering). For this, C_{pm} incorporates the variation in the method with respect to the target value and the specification limits preset by the analyst/customer [26]. This index conveys critical information regarding whether a method (or process) is capable of reproducing items satisfying a requirement that would be preset by the analyst [30]. If the prescribed minimum capability fails to be met, the method is considered incapable.

To calculate the C_{pm}, the method mean and variability must be estimated relative to the method target and specification limits [27]. In our case, the target value T corresponds to standard solution concentration (T = 100 U/mL). The analysis of the measurements during this period shows the accuracy; the average error between

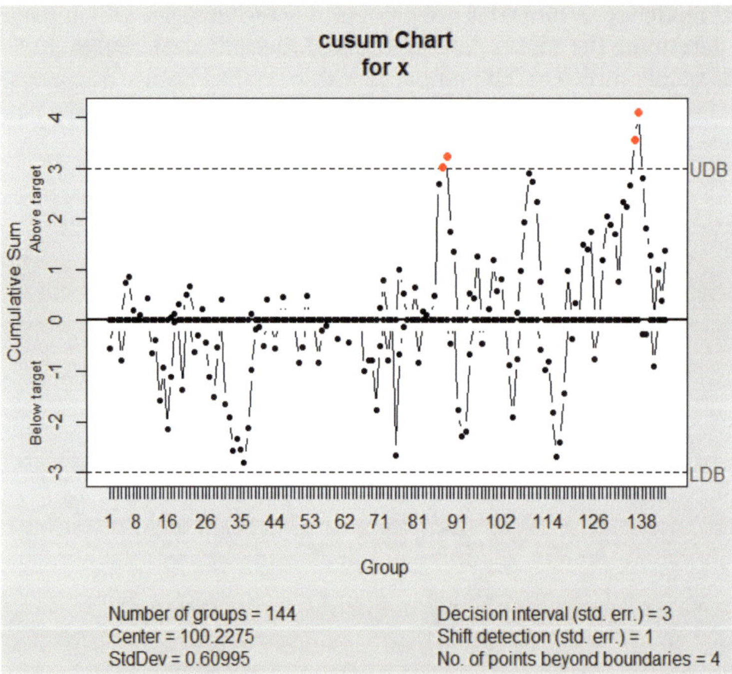

Figure 8.
Cusum control chart for HPLC method. The number of samples beyond limits was four for shift detection fixed at ±1 SD with an upward tendency. The cause of this alteration is unknown.

USL-LSL	C_{pm}	Co	C_{pk}	Co	Method capability
±3SD	0.95	1.08	0.79	1.095	Inadequate
±2.5%	1.30	1.08	1.26	1.095	Capable
±3%	1.56	1.44	1.54	1.45	Satisfactory
±5%	2.60	2.16	2.65	2.18	Super

Table 2.
C_{pk} and C_{pm} values as a function of the specification limit (USL-LSL). The method capability is based on the critical values (Co).

the mean and target concentration $(\mu - T)^2$ was 0.052 U/mL, whereas the precision calculated from the MR-chart was 0.36 U/mL. The overall uncertainty, calculated as the sum of the uncertainty of each component's contribution (precision and accuracy), was 0.41 U/mL. The expanded uncertainty was 0.82 U/mL, using a coverage factor of 2. The calculated concentration is thus 100 ± 0.82 U/mL.

If the specification limits are set at ±5% of the T value, according to Bouabidi et al. [8] criteria, the C_{pm} index was 2.60 (n = 100) with a lower confidence bound C on C_{pm} (i.e., the value used to measure method capability) of 2.48 (**Figure 9**). We therefore conclude that the true value of the method capability C_{pm} is no less than 2.48 with a 95% level of confidence. This result indicates that the method is "super" ($C_{pm} > 2.0$), and no further stringent precision control is required.

If the specification limits are reduced to ±3% of T value, the C_{pm} was 1.56 with a lower 95% confidence limit of 1.47. This result implies that the method is considered "satisfactory" ($1.33 < C_{pm} < 1.50$). The method is inadequate for specification limits lower than ±2% of the T value, since the lower 95% confidence limit for the C_{pm} is less than 1. A similar result was obtained when the specification limits and the control limits were of equal width. Thus, the number of observations out of specification in the method was zero when the specification limits are greater than ±3% of the T value, and the proportion of nonconforming results was less than 1, giving a process yield of 100%. When the reference limits and the specification

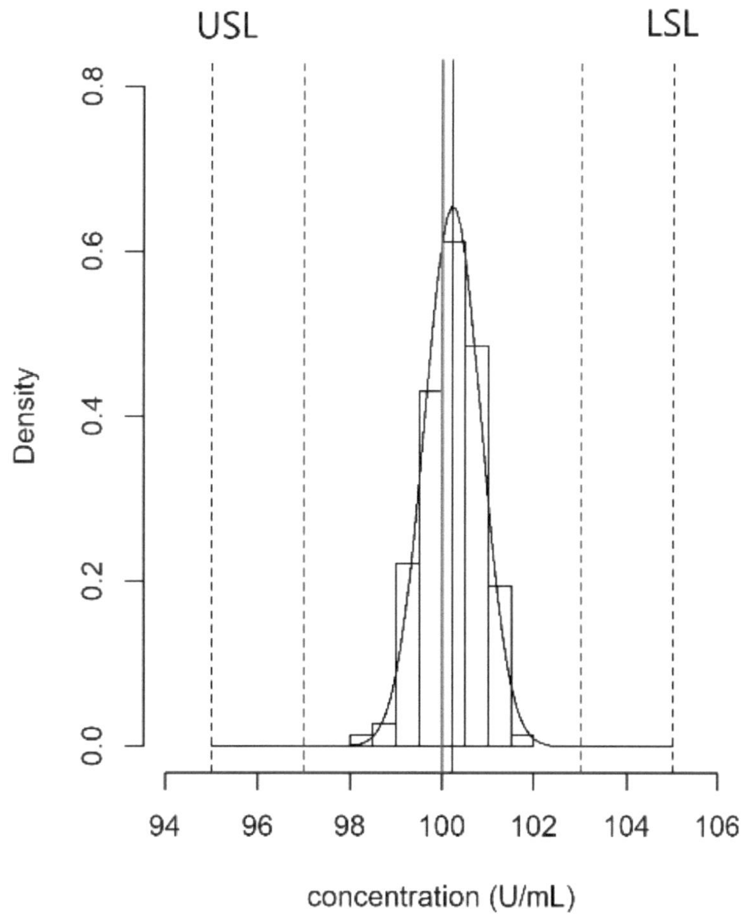

Figure 9.
The black-dashed line shows the specified limits (USL and LSL) established at ±5% of T value, whereas the blue-dashed line corresponds to ±3%. The red line is the T value.

limits are of equal width, only 0.27% of the expected observations will be out of specification in the long term, and the process yield is 100%.

The results obtained showed that we cannot use the control limits as specification limits, since the method is considered "inadequate" according to the criteria proposed by Pearn and Shu [24]. However, the control limits can be used in the development of the specification limits. In this example, a level of variability ±2.5% of T can be enough to declare the method capable ($1.0 < C_{pm} < 1.33$), as seen in **Table 2**.

All these aspects were also analyzed using the C_{pk} index. The results are summarized in **Table 2**. To determine if the method meets the capability requirement, we must calculate the critical value Co for C_{pk} based on α risk, sample size, and C value (i.e., the required method capability) [25]. We find the critical value Co = 1.081, based on C = 1.0, α = 0.05, and sample size n = 200. When the specification limits are set at ±2.5% of T value, the estimated C_{pk} value is 1.30 and exceeds the critical value of 1.081, demonstrating that the method meets the capability requirements. Furthermore, if the limits are increased to ±5% of the T value, the C_{pk} increases to 2.65 with a critical value of 2.18 (based on C = 2.00; α = 0.05; n = 200), indicating that the capability is "super." The obtained quality requirements were similar in both cases.

The use of C_{pk} is clearly preferable when the limits are not equidistant, whereas the C_{pm} index can be overly conservative in this scenario. In our case, the target is at the center of the specification range, and if the aim of our method is to achieve a measure close to the target value with minimum variation, then C_{pm} is the most sensitive capability index. Given two analytical methods with different performances (i.e., precision and accuracy) and the same method departure, a simple comparison between both C_{pm} is sufficient to select the better, although similar results were obtained with the C_{pk} index. This fact was analyzed by Oliva and Llabrés [29].

4. Conclusions

The traditional x-chart and moving range chart represent the first option to monitor the analytical method or process over a long time. The EWAMA and Cusum are two good alternatives in those situations where it is important to detect small process shifts. The capability indices are calculated to evaluate whether the process or analytical method under study is able to provide sufficient conforming results when it is operating under statistical control. To calculate the PCIs, it is necessary to know the actual performance of the process or analytical method (using the control chart) and the customer requirements in terms of specification limits. The specification limits should be determined externally from previous knowledge of inherent process variability. However, different criteria have been proposed to fix these limits. The C_{pm} and C_{pk} indices were used as part of the control strategy. C_{pk} is the best option because it is not dependent on the process or method being centered. However, C_{pm} is more sensitive to departure from the method target than C_{pk} is. Independent of the criteria used to establish the specification limits, computation of the capability indices depends on the analyzed response, and their application is limited to each particular situation and is not general.

Acknowledgements

The authors wish to thank Dr. José B. Fariña from University of La Laguna, Tenerife, Spain, for providing the compaction process data. This research was financed by Instituto de Salud Carlos III, Ministerio de Ciencia, and Innovación y Universidades y FEDER as part of Project PI18/01380.

Conflict of interest

The authors declare no conflict of interest.

Author details

Alexis Oliva* and Matías Llabrés
Department of Chemical Engineering and Pharmaceutical Technology, School of Pharmacy, University of La Laguna, La Laguna, Tenerife, Spain

*Address all correspondence to: amoliva@ull.es

References

[1] Montgomery D. Introduction to Statistical Quality Control. 6th ed. New York, USA: John Wiley & Sons, Inc; 2009. ISBN: 978-0-470-16992-6

[2] Roberts S. Control chart tests based on geometric moving averages. Technometrics. 1959;**42**:97-102

[3] Crowder S. Design of exponentially weighted moving average schemes. Journal of Quality Technology. 1989;**21**: 155-162

[4] Lucas J, Saccucci M. Exponentially weighted moving average control schemes: Properties and enhancements. Technometrics. 1990;**32**:1-29

[5] Lucas J. The design and use of V-mask control schemes. Journal of Quality Technology. 1976;**8**:1-12

[6] Hawkins D, Olwell D. Cumulative Sum Charts and Charting for Quality Improvement. New York, NY: Springer Verlag; 1998. ISBN: 978-1-4612-1686-5

[7] Oliva A, Llabrés M, Fariña J. Capability measurement of size-exclusion chromatography with a light-scattering detection method in a stability study of bevacizumab using the process capability indices. Journal of Chromatography A. 2014;**1153**:89-98. DOI: 10.1016/j.chroma.2014.04.027

[8] Bouabidi A, Ziemons E, Marini R, Hubert C, Talbi M, Bouklouze A, et al. Useful capability indices in the framework of analytical methods validation. Analytica Chimica Acta. 2012;**714**:47-56. DOI: 10.1016/i. aca.2011.11.019

[9] Dejaegher B, Jimidar M, De Smet M, Cockaerts P, Smeyers-Verbeke J, Van der Heyden Y. Improving method capability of a drug substance HPLC assay. Journal of Pharmaceutical and Biomedical Analysis. 2006;**42**:155-170. DOI: 10.1016/j.jpba.2006.01.001

[10] Kamberi M, García H, Ferer DP, Rapoza R. Setting acceptance criteria for validation of analytical methods of drug eluting stents: Minimum requirements for analytical variability. European Journal of Pharmaceutical Sciences. 2011;**42**:230-237. DOI: 10.1016/j. ejps.2010.11.013

[11] Raske C, Bennett T, Goodberlet S. Risk-based analytical method transfer: Application to large multi-product transfers. Analytical Chemistry. 2010; **82**:5932-5936. DOI: 10.1021/ac1008892

[12] Oliva A, Llabrés M, Fariña J. Development of two high-performance liquid chromatographic methods for the analysis and characterization of insulin and its degradation products in pharmaceutical preparations. Journal of Chromatography B. 2000;**749**:25-34. DOI: 10.1016/s0378-4347(00)00374-1

[13] Reynolds M, Stoumbos Z. Monitoring the process mean and variance using individual observations and variable sampling intervals. Journal of Quality Technology. 2001;**33**: 181-205. DOI: 10.1080/00224065.2001. 11980066

[14] Batson R, Jeong Y, Fonseca D, Ray P. Control charts for monitoring field failure data. Quality and Reliability Engineering International. 2006;**22**: 733-755. DOI: 10.1002/qre.725

[15] R Core Team. R: A Language and Environment for Statistical Computing. Vienna, Austria: R Foundation for Statistical Computing; 2018. Available from: https://www.R-project.org/

[16] USP General Chapter <1216> Tablet friability. In: USP37-NF32. Rockville: The United States

Pharmacopeial Convention; 2014. ISBN: 978445237700

[17] Maroto A, Boqué R, Riu J, Rius F. Estimation of measurement uncertainty by using regression techniques and spiked samples. Analytica Chimica Acta. 2001;**446**:131-143. DOI: 10.1016/S0003-2670(01)00842-X

[18] de Vargas V, Dias Lopes L, Mendonca-Souza A. Comparative study of the performance of the CuSum and EWMA control charts. Computers and Industrial Engineering. 2004;**46**: 707-724. DOI: 10.1016/j.cie.2004.05.025

[19] Oliva A, Llabrés M. Validation of a size-exclusion chromatography method for bevacizumab quantitation in pharmaceutical preparations: Application in a biosimilar study. Separations. 2019;**6**:43-55. DOI: 10.3390/separations6030043

[20] Kotz S, Johnson N. Process capability indices—A review 1992-2000. Journal of Quality Technology. 2002;**34**:2-19. DOI: 10.1080/00224065.2002.11980119

[21] Spiring F, Leung B, Cheng S, Yeung A. A bibliography of process capability papers. Quality and Reliability Engineering International. 2003;**19**:445-460. DOI: 10.1002/qre.538

[22] Boyles R. The Taguchi capability index. Journal of Quality Technology. 1991;**23**:17-26. DOI: 10.1080/00224065.1991.11979279

[23] Wu C, Pearn W, Kotz S. An overview of theory and practice on process capability indices for quality assurance. International Journal of Production Economics. 2009;**117**: 338-359. DOI: 10.1016/j.ijpe.2008.11.008

[24] Pearn W, Shu M. Lower confidence bounds with sample size information for Cpm applied to production yield assurance. International Journal of Production Research. 2003;**41**: 3581-3599. DOI: 10.1080/0020754031000138349

[25] Pearn W, Chen K. Making decisions in assessing process capability index Cpk. Quality and Reliability Engineering International. 1999;**15**:321-326. DOI: 10.1002/ (SICI)1099-1638

[26] Pearn W, Shu M, Hsu B. Testing process capability based on Cpm in the presence of random measurement errors. Journal of Applied Statistics. 2005;**32**:1003-1024. DOI: 10.1080/02664760500164951

[27] English J, Taylor G. Process capability analysis—A robustness study. International Journal of Production Research. 1993;**31**:1621-1635. DOI: 10.1080/00207549308956813

[28] Cano E, Moguerza J, Corcoba M. Quality Control with R: An ISO Standards Approach. Switzerland: Springer; 2015. ISBN: 978-3-319-24046-6

[29] Oliva A, Llabrés M. Application of capability indices and control charts in the analytical method control strategy. Journal of Separation Science. 2017;**40**: 3046-3053. DOI: 10.1002/jssc.201700173

[30] Shu B, Shu M, Pearn W. Measuring process capability based on Cpmk with gauge measurement errors. Quality and Reliability Engineering International. 2007;**23**:597-614. DOI: 10.1002/qre.836